ALCHEMY ASTROLOGY

Lost Key To The Philosopher's Stone

Astrology Handbook for Practical Laboratory Use

This book is intended for informational purposes. The author, agent, or the publisher will not be held accountable for the use or misuse of the information in this book.

Copyright © 2011 Timothy A. Wilkerson
All rights reserved.

ISBN: 978-1-257-98077-2

No part of this book may be reproduced or utilized in any form, or by any means, electronic or mechanical, including photocopying, recording or by any information storage and retrieval system, without permission in writing from the author Tim Wilkerson. The blank charts and diagrams in the Appendix may be copied without permission for personal or instructional use only. Reviewers may quote brief passages.

Star information regarding co-culmination, co-rising and co-setting positions, and general influences, provided by permission from Bernadette Brady author of *Star and Planet Combinations*.
www.bernadettebrady.com

Cover illustration by Timothy A. Wilkerson.

Background space photographs used for the cover art were permissively obtained from NASA websites, as they are free for use in the public domain. Application of these images does not denote, nor imply, any form of endorsement for this publication by NASA.

Printed and bound in the United States of America by Lulu.com.

Contact the author: tim@alchemyastrology.com

Table of Contents

Dedication	i
Acknowledgements	i
Forward by Dennis William Hauck	ii
Bio: Dennis William Hauck	iii
Preface	iv
Introduction	v
Guidelines	vi
CHAPTER ONE – THE BIG PICTURE	**1**
1. Alchemy & Science	*1*
2. Astrology & Astronomy	*2*
3. The Philosopher's Stone	*4*
4. The Vegetable Stone	*5*
5. Why Astrology Is The Lost Key	*5*
6. Microbes & The Philosopher's Stone	*6*
7. The Adept Practitioner	*9*
CHAPTER TWO – ASTROLOGICAL CHARTING	**11**
1. How To Make A Basic Tropical Zodiac Astrology Chart	*11*
2. Placing The Planets	*13*
3. The Aspects	*14*
4. How To Make A Sidereal Zodiac Astrology Chart	*16*
CHAPTER THREE – THE FIVE LAWS	**19**
The Five Laws	*19*
1. Law One - Polarity Of The Planets	*20*
2. Law Two - Influences Of The Zodiac	*22*
3. Law Three - Duration Of The Influences	*23*
4. Law Four – The Four Elements	*24*
5. Law Five - The Aspects	*26*
6. The Total Count	*27*
Conclusion	*28*
CHAPTER FOUR - INFLUENCES	**29**
1. General Influences	*29*
2. Influences Of The Planets	*29*
3. Influences Of The Major Aspects	*31*
4. Influences & Lab Processes	*31*
5. Influence With The Time Of Day	*33*
6. Influence Of The Moon	*35*
7. Other Influences	*36*
CHAPTER FIVE – INTERPRETING THE CHART	**37**
Conclusion	*39*

CHAPTER SIX – THE STARS   41
1. The Stars of the Ancients'   41
2. Star Influences List   44
3. Interpreting A Chart Including The Fixed Stars   45
Conclusion   46
4. Star Paran Tables   47

CHAPTER SEVEN – RELATING TO MATTER   51
1. Herbs To The Planets   51
2. Metals To The Planets   52
3. Body To The Zodiac & Planets   52
4. Body To The Stars   53

CHAPTER EIGHT - 2012   55

Appendix A   61
1. Planets – Scientific Stats   61
2. Stars – Date of Record, Distance, & Type   61
3. Stars – Names & Scientific Spectral Type   62
4. Stars – Spectral Type & Associated Temperature Ranges   63
5. Constellations Of Stars   64

Appendix B   65
1. Blank Astrology Chart   66
2. Transparency   68
3. Blank Earth Centered Orbital Charts   70
4. Basic Lab Form   74
5. Quick Chart – Four Day   76
6. Salts - Calcining Grayscale Reference   78

Appendix C   80
1. List of Website Resources for Further Study   80
   a. Alchemy   80
   b. Astronomy   80
   c. Astrology   80
   d. Documentaries   80
   e. Microbial Research   81
2. Recommended Books   81

Works Cited   82

Dedication

This handbook is dedicated

to all the faithful alchemists

whose research, writings, teachings, and experiments,

have kept alchemical pursuits alive

over the last two millennia.

Acknowledgements

I'd like to acknowledge the late Frater Albertus for teaching me the basics of Laboratory Herbal Alchemy, and for the subtle clues he provided. He inspired a deeper understanding of how astrology relates to myself, the world, alchemy, and the Philosopher's Stone (a.k.a. the Sorcerer's Stone).

I want to thank Dennis William Hauck for the opportunity to speak on this subject in front of an extremely interested audience, and figuratively pointing at the Moon while suggesting I write a book.

I relay my deep appreciation to Bernadette Brady for permission to reprint a portion of the Paran Tables from her amazing book, *Star and Planet Combinations*.

Special thanks to Teri Owens for her patience and support, over the two years it took me to research and write this publication, and for her assistance with content, formatting, and grammar.

Also, thanks to Peg Aloi, for her professional editing expertise, and Alan Salmi, for references to new astrology research.

Peace, love, time and space, abundant,
Tim Wilkerson

Forward by Dennis William Hauck

I first met Tim Wilkerson at the first International Alchemy Conference in 2007. As director of the conference, I had invited former students of Frater Albertus (Dr. Albert Reidel) to be our guests of honor. Dr. Reidel was a pioneer of American alchemy and founded the Paracelsus Research Society in Salt Lake City. That group evolved into Paracelsus College, where Reidel taught alchemy based on Paracelsian concepts, such as the Three Essentials (Body, Soul, and Spirit) and spagyric techniques of the separation and recombination of purified plant essences to create alchemical medicines. Reidel's <u>Alchemist's Handbook: Manual for Practical Laboratory Alchemy</u> is still the standard textbook for plant alchemists.

Tim Wilkerson was one of a handful of alchemists who had the opportunity to study with Dr. Reidel and graduated the Prima class at Paracelsus College in 1984. But I soon discovered that Tim was more than an alchemist. He was a true Renaissance man: noted musician, artist, author, astrologer, lecturer, teacher, medicinal herbalist, and entrepreneur. Tim's breadth of knowledge and experience was obvious from our first conversations, and I invited him to be a speaker at the 2008 International Alchemy Conference.

Tim's lecture on laboratory astrology was extremely well received. He showed how to do a basic astrology chart in relation to alchemical laboratory work. He explained how astrology not only gives us indications for the best time to start an experiment, but which lab processes will complete more successfully and efficiently, in a given time frame.

After the lecture, I realized what breakthrough work Tim was doing. Alchemy has been associated with astrology since ancient Egypt, but Tim modernized those ancient theories with practical applications based on his own meticulous research into the connection between the stars and alchemical experiments. His focus on sidereal influences took the work in laboratory astrology to a new level. Previous work by Dr. Reidel and Swiss alchemist Manfred Junius had focused on planetary effects on the outcome of alchemical operations, and alchemists had been timing their lab work using planetary hours charts that were hundreds of years old.

Tim was very modest about the importance of his work, but I urged him to organize his research into a book. He resisted at first, as he considered all the work involved in such a project. But then his imagination took over, the stars inside him started shining, and I could see their glimmer in his eyes. The manifestation of that light lies before you.

Bio: Dennis William Hauck

Dennis William Hauck is an author, consultant, and lecturer working to facilitate personal and planetary transformation through the application of the ancient principles of alchemy. He writes and lectures on the universal principles of physical, psychological, and spiritual perfection to a wide variety of audiences that range from scientists and business leaders to religious and New Age groups. Hauck's interest in alchemy began while he was still in graduate school at the University of Vienna, and was initiated into the craft by a practicing alchemist in nearby Prague.

Hauck has since translated a number of important alchemy manuscripts dating back to the fourteenth century and has published dozen books on the subject. His bestselling The Complete Idiot's Guide to Alchemy (Penguin Alpha 2008) is considered the best introduction to alchemy available today, and his The Emerald Tablet: Alchemy for Personal Transformation (Penguin 1999) presents Hauck's original research about the mysterious artifact that inspired over 3,500 years of alchemy. His Sorcerer's Stone: A Beginner's Guide to Alchemy (Penguin Citadel 2004) is an entertaining introduction to practical and spiritual alchemy.

Hauck holds lectures and workshops throughout the world on the various aspects of practical, mental, and spiritual alchemy. He is founder of the International Alchemy Conference (*www.AlchemyConference.com*), an instructor in alchemy (*www.AlchemyStudy.com*) and is president of the International Alchemy Guild (*www.AlchemyGuild.org*). His website is *www.DWHauck.com*.

Preface

Most of the ideology in this handbook was passed on to me from my alchemy instructor by word-of-mouth, in the oral tradition. Since 1984, I've used my experience in the school laboratory, along with my class notes on astrology, to conduct my own experiments in order to confirm what I had been taught. In my little home laboratory, I continued my investigations by making plant remedies for myself, and my family. I always found great delight observing how my attempts to repeat, and exact, an experiment were consistently altered by the astronomical positions of the planets. This has proven to me that astrology has foundation in the physical world, and I feel compelled to give you the opportunity to determine the value of this instruction for yourself.

To encourage your intuition regarding the workings of astrology in the physical world, I have attempted to morph science with alchemy, and astronomy with astrology. I feel my research of scientific data, theories, and recent discoveries, lend foundation to my concepts. I must admit, a good portion of the scientific information cited herein is so new that it hasn't found its way onto bookshelves yet. It can only be found in the form of film documentaries, publicly funded scientific websites, university level research papers, current news or magazine articles, and abstracts from commercial sources. Therefore, I must warn you, some of my references may still be considered controversial by mainstream scientists.

Alchemy spans many different aspects of study, but my aim here is to simply increase your understanding of astrology as it applies to alchemical theory. Furthermore, my ideas about the reality of the Philosopher's Stone are an extrapolation of my experience and study of alchemical writings, with a dash of contemporary scientific knowledge thrown in. So, please, do not automatically believe what is stated in my writings, instead, carefully apply this information to prove it to yourself. It can be quite a phenomenal journey.

A reference list of related citations, and internet resources, can be found in Appendix C.

"Do not fear to be eccentric in opinion, for every opinion now accepted was once eccentric."
-- Bertrand Russell

Introduction

Simply defined, Practical Alchemy is the study of natural laws as observed in a controlled laboratory environment. It includes the search for, and concentration of, the first creative energy believed responsible for bringing life to our planet Earth. The alchemists call this the Prima Materia and it is believed to be a universal form of matter/energy. When we learn to focus, and concentrate this subtle energy into a purified substance, we are working with the Philosopher's Stone. Astrology is a tool of observation, and reliance, used to deduce the natural laws of alchemy.

As it's practiced today, Practical Alchemy is predominantly based on the written research records of Alchemists born in the last eighteen hundred years. A great volume of earlier information was lost when war brought destruction to the vast libraries of past civilizations. Alchemical studies currently available are once again nearing the grandeur of intelligence that was available in ancient times. This handbook is an example of the wisdom passed down from teacher to student, combined with my personal research, and the expertise of others devoted to alchemy and astrology.

The specific goal of this book is to help alchemists determine the best time to start a laboratory project by determining how many influences from the cosmos are dynamically focused on our planet. The general goal is to provide a tool to predict energies available for any important activity, or event. This knowledge is basic, and is written as a tool for contemplation. As you watch the world, and practice alchemical astrology, the single word descriptions will begin to provide insight to the kinds of Energy that can be directed at us. Observation of astronomical alignments as compared to earthquakes (USGS, 2010), volcanic eruptions (Smithsonian Institute, 2010), and severe weather conditions, can further delineate your understanding of the physically observed influences.

Experience has shown me that when astrological conditions are favorable my experiments proceed in an accelerated fashion. The resultant product is more potent, and laboratory accidents are unlikely. There may be occasions when you can't wait for the best astrological situation. Perhaps you need to prepare a remedy for someone who is too sick to suffer any postponement. Your alchemical preparation may not be as strong as it could be, but any neglect to act would be detrimental to the patient. In those situations it is important to know how and when a problem is likely to arise, so that it can be avoided or compensated. You can make the best of the most advantageous methods available, and choose a different strategy to complete the task at hand, when you consider the cosmic astrological influences.

Guidelines

Alchemists from the fifteenth century through to contemporary times used seasonally aligned Tropical Astrology charts. For decades I consulted this system to predict a good time to start my plant experiments. I found it surprisingly helpful, and relevant to my observations, but sometimes there were holes in my chart interpretation that left me scratching my head. Then, I discovered Sidereal Astrology (Blair-Ewart, 1999), a star aligned system that is astronomically accurate. A short time later the unexplained charting gaps seemed to disappear, which bolstered my convictions further. If you're familiar with your tropical birth chart, try comparing it to a sidereal version, and you may be fascinated to find some unrealized facets of yourself. I am biased to this sidereal Zodiac, but I will present you with the basics, so that you can determine which of these two systems, tropical or sidereal, if not parts of both, you prefer to use in your work.

I look ahead for the best day to start the lab work, on dates when I have the free time necessary to govern and maintain a high level of attention. A week-long vacation, or a period when I know I won't have to leave the house for a day or two, is perfect. This uninterrupted time is optimal because I never leave my lab running unattended, and I would rather not have to start and restart the experiment. Using the ephemeris, I create quick charts (Appendix B) to find the future active and passive totals during planned my time off. I evaluate the days with the best positive numbers, as calculated using the Five Laws. It is up to each alchemist (or scientist) to carefully determine, and record, the particulars of how this information is helpful in their laboratory work. (A list of recommended books regarding the practical laboratory work, astronomy, and astrology, can be found in the Appendix C.)

Unless otherwise specified, when I talk about placing the planets in a chart, I'm referring to the seven major cosmic bodies in our solar system. These are the celestial objects that can be seen with the naked eye; Sun, Moon, Mercury, Venus, Mars, Jupiter, and Saturn. Alchemically speaking, they are hypothesized to be the reason we have: seven musical notes, seven basic colors, seven geographical Tectonic Plates, and seven years for our bodies to complete total cellular renewal. Also, human awareness and intellectual growth, has a seven-year cycle. At ages 7, 14, 21, 28, 35, 42, etc., we can experience shifts in our view of ourselves, and our environment.

Records of the cosmic influences attributed to the stars date back thousands of years, and it is to this fact that I allude to when I reference the Ancient Alchemists (merlynne6, 2009). In my perspective, they are the ones who conceived how to: make ceramics circa 29,000 BC, purify metals circa 3,000 BC, and refine glasswork circa 1,000 BC. Concrete was formulated in 20 BC, and was probably created by an alchemist. Consequentially, such discoveries as these prior to 200 AD, are likely to have been found using clues given by relation to the actual position of the Stars.

I have chosen two stars per sign. The ones listed have been selected because they are more likely to have been used by ancient achemists. Most are listed in books as the "Brightest Stars" and are written in the histories of ancient peoples (Gregory Cajete, 2000), or painted on the walls of prehistorically inhabited caves (Rappenglueck, 2006). I've also included stars from constellations named after animals mentioned in Alchemical texts, such as, Leo the Lion, and Draco the Dragon (Hauck, 2008). A good example is Vega the Vulture, who is associated with the alchemical process of sublimation. Some selections are associated to records of alchemical gold making ranging from 1382 to 1782 AD (Kollerstrom). There are many other stars of interest to investigate but this is a good start.

There are three concepts of astrology omitted from this publication because I have yet to discover any correlation to Practical Alchemy. The twelve houses are not covered because their development is a relatively recent addition. Originally, there were only eight houses that proceeded in a clock-wise direction, and were primarily related to the time of day as applied to personal life cycles (Guinard, 1999). I haven't listed the rather modern particulars about governing planets as related to the zodiac signs because they are linked to the seasons, and I prefer to focus on influences related to the stars themselves. Rising signs are also thought to be decisive in today's contemporary theories. Although they are computed using sidereal astrology I haven't found any unique significance to their consideration in the laboratory. You may find houses, Zodiacal governing planets, and rising signs, useful in your birth chart (McEvers & March, 1976), and I'll let you know if I find anything new about them for practical purposes.

Please, for your safety, and the safety of your family and neighbors, always have a fire extinguisher in your laboratory, and never leave it running while unattended! Especially when you have to work under difficult astrological, or environmental, influences.

As always, please do not ingest 'any medicine' without sound medical monitoring!

CHAPTER ONE – THE BIG PICTURE

1. Alchemy & Science

Alchemists have their tests, but they are dissimilar to the attempts of modern scientists to find intellectually exclusive and 'absolute' comprehension of the great mysteries in our Universe. Alchemists are endowed with a sense of liberation because they are exclusive of the pressure to empirically prove their results as scientific fact. To be an alchemist you have to open your mind and go beyond the contemporary intellectual consensus to find answers. The realizations provided by this way of thinking, enlighten the alchemist who is enamored with the natural laws.

Alchemical properties are recognized when a 'result or 'reaction' occurs that baffles modern scientists. Science can't explain exactly how the symbiotic microbes in bees' digestive systems function to create honey, so it can be defined as an alchemical reaction. As another example, consider how our bodies create calcium for bones and muscles when our diets are low in calcium. If we ingest the herb horsetail, which is high in silica, along with a food high in carbon, our bodies can alchemically transform these two substances. Their atomic weights combine to total the weight of calcium. This can be defined as an alchemical law and is known as Biological Transmutation (Staelin, 2006). In other words, alchemy doesn't put a great deal of importance on exactly the how to make honey, or calcium, but instead emphasizes the usefulness of the function of natural laws that create them.

The laws of alchemy, as used in the laboratory, can be applied to modern technology. When Isaac Newton applied the alchemical theory, 'As above, so below, as below, so above', to a falling object, his investigations led him to realized that the same laws that govern an apple's descent from a tree, could also be applied to understanding the cosmos (Hatch, 1998). Isaac Newton was an alchemist, who also became a scientist (Hauck D.). A recent example, availing natural alchemical laws to science, is when bacteria are cultured to process garbage and oil spills. Wineries are another good illustration of Science helping to improve on the natural laws that govern their employment of microbial yeasts. Premium cultures secured from their fermentation vats are reused to seed the proceeding batch, and have on occasion been passed down through family lines for centuries. Often these hybrid microbes are unique when compared to those found in nature due to the alchemical efforts of the wine makers. Decidedly, it is the level of artistry attained by those who create wines that prove them to be fine products, but today, science provides great tools and instruments that aid in identifying, protecting and promoting, these beneficial yeast communities. Perfection of the arts of fermentation and distillation is a very desirable talent in the realm of the alchemist.

ALCHEMY ASTROLOGY

"I am enough of an artist to draw freely upon my imagination. Imagination is more important than knowledge. Knowledge is limited. Imagination encircles the world."

-- Albert Einstein

2. Astrology & Astronomy

Astrology is a tool that we can use to monitor natural events and changes in our environment. It is a time-honored way of recording Astronomical conditions in the form of a chart that is easy to understand. We don't have any precise scientific way to detect how the gentle push and pull of our Sun, and fellow planets, affect our fluid and ever-changing magnetosphere, nor how these fluctuations could affect our bio-sphere. What modern astronomers can detect though, is that when distant stars wobble, it proves that they have a planet in orbit around them. As planets spin and revolve in their orbit gravity visibly tugs on their host stars, and our Earth and Sun interact exactly the same way. Ultimately, human intuition coupled with astronomy, will provide confirmation that what we see in the lab is influenced by these cosmic interactions. Effects that would normally go unnoticed if it weren't for the astrological method of charting, and displaying current conditions, will be unveiled.

Ancient peoples originally based their knowledge of astrology on the actual positions of the constellations. In consideration of past observations they knew that when certain stars align with a planet their energies were focused and amplified in predictable way. Around 200 AD this basic theory of astrology was collected by scholars studying in Egypt, and presented to Greek society. Unfortunately, the fact that the stars appear to move in relation to the Earth was lost. Due to the wobble of the Earth's axis the stars progress across the sky at a rate of about one and a half degrees every seventy years, and the Egyptians would consistently calculate adjustments to their charts.

When astrology arrived in Greece the spring equinox was aligned with Aries, and contemporary astrologers are still using this format. They move the zodiac to align it with the spring equinox. This means that since 200 AD the energy of the stars has been ignored. Coincidently, reports of gold making have gradually faded from history. As the stars progressed, the influence of cosmic energies became less apparent, and gradually fewer alchemists were able to produce the Stone. Over time, the key to creating the Stone slipped away from human understanding, and became a rumor of the past. For astrological information to be useful we must be able to accurately predict stellar influences and properly associate them with greatest volume of past observations.

Astrological observations were compiled through the experiences of ancient people who had to be acutely aware of their environment in order to survive. The height of the Moon in alignment with the Sun often marked tsunami and earthquake propensities (Sanders, 2009). The winds, the weather, the return of plants in the spring, the changing length of daylight, and planting and harvesting times, were all encompassing events that if missed, could bring starvation. The Moon was observed to affect the feeding habits and breading activities of prey (Hansford, 2007), and the height of the Sun marked the migratory movements of fowl, allowing hunters and fishers to plan ahead. All of these aspects were viewed as interacting with each other. Energies from the star constellations not only set the stage where everything takes place, they were understood to radiate the very fuel of life itself.

A good example of how astronomical conditions can impact us is evident when we strive to understand why our Earth is experiencing a significant rise in global temperature. It's been reported that other planets in our solar system show signs of global warming (Than, 2007). This is because, between 2002 and 2008, sunspots have gradually diminished to the point where, between 2008 and 2010, our solar system experienced an unprecedented lack of sunspots. This is unusual when compared to our admittedly limited record of past cycles. Like the Earth's magneto/bio-sphere, which protects us from harmful solar radiation, the Sun has a helio-sphere, which encompasses and protects our entire solar system from stellar radiation. It is stronger when sunspot activity is high, but currently it is weaker than usual. This helio-sphere shields us, but when the energy of the Sun is low, it allows more cosmic energy to reach the planets. It is suspected that this increase of cosmic radiation is the reason for global warming in the solar system, including our Earth.

I imagine planets traveling in our solar system creating wakes like big boats navigating a river, or a cargo jet flying through the air. I construct this mental imagery by adapting the following scientific considerations. Astronomers and mathematicians have located Lagrange Coordinates where space vehicles can park in a neutral gravitational area (Boyd, 2009). It's easy to envision a similarity to a harbor, or a gently flowing spot on a river, where space vehicles can shore-up with minimal flight adjustments and fuel expenditure. Science also knows that as the current of the Solar Wind passes by the almost gravity-free Moon, defined areas inhabited by magnetic rocks experience an amplification of their overall magnetic field(Lin, et al., 1998). I wonder at comparing this to a large volume of high velocity water accelerating over the edge of a narrow waterfall that pours onto the Earth. After all, it is common knowledge that the Moon and the Sun's gravitational interactions obviously affect the ocean as observed in the different tidal cycles they create. Earth is orbiting through these water-like electromagnetic and gravitational fields within our solar system.

Similar to our earthly weather prediction systems, Scientists will one day understand the mathematics involved in predicting the heliocentric gravitational currents of our solar system. Wave theory, using mathematical logarithms, can explain rouge waves like those generated near Cape Horn. Do these equations relate to gravitational effects in our solar system? When small electro-magnetic flashes are detected at fault zones just before an earthquake (SIUC, 2010) are these charges triggered by planetary geometry? We know that heat is absorbed by the seas and is collected and released in the form of hurricanes, thus balancing Earth's global atmospheric temperature. Does our galaxy behave this way, collecting and directing excess energies to remedy an imbalance? Throughout the Universe there are rhythms like these, flowing between light and dark matter, that may act similarly to the circulation of our seas. When we can scientifically create a global map of our current gravitational magnetosphere, and watch how it changes when energies triggered by the planets move into different angles to the Earth, we could make a useful solar weather forecast. Until then, astronomy and ancient alchemical astrology, are our only clues. They are our guide-map to furnish the advantage of prevailing currents on our journey through the cosmos.

"Look deep into Nature, and then you will understand everything better."
-- Albert Einstein

3. The Philosopher's Stone

There are two types of Stone that alchemists strive to create. One is the Vegetable Stone made from plant matter, and the other is the Mineral Stone made from rocks, or metal ore, also known as the Philosopher's Stone. The Vegetable Stone can be made in less than one year, the Philosopher's Stone can take decades. Both are processes that raise the vibratory rate of the material you are processing, in order to match the high vibration of human beings, thus making it a more evolved remedy. The test to prove that the Vegetable Stone has been correctly prepared, is to suspend it in an Herbal maceration (or tea) so that it may draw the purified medicinal constituents to the surface, where it can be collected, and applied. During this process, the Vegetable Stone will enlarge to a degree, or at least it will maintain its original size. The making of gold, by transmutation, is the test to prove that the Philosopher's Stone is ready to be used as a medicine. If it can't transmute metals, it won't be able to transform the body to good health. Unlike the Vegetable Stone, once it is used the Philosopher's Stone can't be reclaimed.

Related myths describing the Stones' ability to extend life, and transmute base metals have been told for millennia. Whole cities of gold, and fountains of youth, have been entertained in the ancient stories of several cultures around the globe. Many people, in the last thousand years, convinced so strongly in the truth of these historical accounts, risked their lives to find and prove them. I also believe the Philosopher's Stone is real, (Kollerstrom) and that it provides important evidence for understanding how to cure, and maintain, our physical health.

Today, there are some alchemists who believe that the Philosopher's Stone emits a primal Energy field similar to the kind of radiation that is used for medical purposes. They think this because, as it is used to renew health, the patient shows symptoms similar to radiation poisoning: hair, fingernail, and tooth loss, are sometimes described when a person imbibes the Stone. Alchemically prepared herbs are required to aid the patients' symptoms while these cells re-grow. After the process is complete the patient appears healthy and even youthful.

This handbook will theorize why historical claims about the Philosopher's Stone are correct, and what factors should be considered in order to re-create it.

4. The Vegetable Stone

It is strongly advised that alchemists begin their study with the Herbal Work before moving on to the greater work of creating the Philosopher's Stone. The herbal work will guide you to the Vegetable Stone, which is the herbal equivalent of the mineral based Philosopher's Stone. The energies of the Vegetable Stone can be used to create spagyrics from herbal emulsions (tea), which is a similar process to transmuting metals with the Philosopher's Stone. It also is a general panacea but is much milder in its effect. It has been proposed by some alchemists that the Vegetable Stone can even be used to create the Philosopher's Stone.

The Herbal Work will give you the lab skills necessary to safely accomplish the more dangerous, and possibly deadly, Mineral Work. With experience and forthright practice, guided by the proper awareness of the astrological influences, you can prepare yourself for the Great Work, and the creation of the Philosopher's Stone. In the meantime the vegetable work will teach you how to make and preserve herbal remedies. If a teacher is available in your location I suggest you inquire about an apprenticeship.

If you wish to help people heal themselves be aware that alchemically prepared herbs and minerals are very potent and should be diluted. Always consult a health care professional before ingesting them. From this work, you may find confidence in your investigation of alchemy's natural laws, and bring greater health to yourself, and your community.

5. Why Astrology Is The Lost Key

With the naked eye we see only a small spectrum of the energy that stars emit, but subtle cosmic influences have been felt by people for eons. Astrology attempts to label and discuss these observed energies from the cosmos. Though astrological theorems are subjective, they are gathered over long periods of time and have a basis in the permanency of the stars. It is the observation of these feelings that were originally used to define the zodiac signs, and delineate the constellations. People who lived on continents divided by the great expanse of our oceans, and had no known contact with each other, primarily came to similar conclusions about the influences of the stars. (Gregory Cajete, 2000)

Astrology gives us clues to energetic phenomenon that can be proven in the sensitivity we carry to our work. Similarly, modern scientists are currently experimenting with how electromagnetic fields affect living organisms. They are finding collaborative evidence in their study of microbes and small marine creatures such as corals and sponges (Hansford, 2007). It is understandable that astronomical emanations can affect our sensitive laboratory experiments when you consider that our Sun also creates electromagnetic energy waves when plasma ejections strike the Earth's protective magnetosphere. Modern scientific methods have similarly detected many other types of subtle energies. Some of these energies are defined as gamma rays, x-rays, radio waves, and intense gravitational fields. The Aurora Borealis is one such product of the Sun's solar radiation.

With the availability of orbital telescopes and spectrum analysis, science has discovered that stars are the source of all metals present in the Universe. As a result scientists speculate that everything made of matter comes from the stars (Geographic, 2010). Inversely, through the use of higher mathematics, astronomers have discovered dark matter (White, 2010) (Chandra, 2006). It appears as the black expanses between all astral bodies, and comprises about one third of the universe. It is reminiscent of what ancient alchemists called the ether, a rarefied element believed to fill the upper regions of space.

Every year there are new discoveries from those deep-space-viewing orbital telescopes, and interplanetary probes, that are changing the way science sees the universe. They are just beginning to understand how objects from space affect each other, our World, and the living beings that dwell here (Junius, 1985). As technology gets more advanced it is probable that many of the facets of alchemy and astrology will be re-discovered. Simply put, especially regarding laboratory processes that take a great deal of time, the astrological conditions can speed up reactions and cultivation. Cosmic influences can also dispel an expected result, thus adding to the importance of astrological predictions.

6. Microbes & The Philosopher's Stone

It is the theory of some scientists that microbes, deposited on Earth by meteors, sparked life on our planet over four billion years ago (NASA, 2000). They survive space travel (Klyce), can thrive on Earth under high pressures and temperatures, and are often found in highly acidic, or very base, environments (Tyson, 2000). They can lay dormant for thousands, and perhaps millions, of years, and can be reactivated under the right conditions (Hanson, 1995). These microbial bacteria are very close to the bottom of the domain tree of life, which is reminiscent of the term *prima materia*, or "first matter", a phrase often used by alchemists. The goal of the alchemist is to capture, refine, propagate, and preserve as a Stone, this Prima Materia so as to prove it in the transmutation of metals, and to enact a universal cure for the human body.

Historically, the Philosopher's Stone is described as a deep ruby red crystal. In nature crystals grow underground when the right environmental conditions are present. Heat, pressure, confinement, and purified minerals in solution, are part of the prerequisites to their development (NatGeo, 2007). These conditions are similar to what the alchemists endeavored to recreate in their laboratories. What they may, or may not, have intellectually realized is that the activity of microbes is what influences natural crystals to grow. The renowned Emerald Tablet, upon which was written the alchemical philosophy of the universe, in base-relief without being engraved, is also representative of the action of microbes to convert a liquid mineral to a crystalline form, as if it were poured into a mold. (Hauck D. W., 2008) Just as microbial yeasts transform vegetable matter into alcohol Spirits, which is a critical process for creating the Vegetable Stone, so might mineral microbes be an integral part of creating the Philosopher's Stone.

Modern science has recently revealed that microbes can induce electrons to move from one substance to another (Amherst, 2001). They have also identified microbes that can eat rock and transform it into crystalline structures comprised of either minerals, or metals, not found in the substances they are consuming. Scientists now find it apparent that the activity of microbial bacteria unearthed in Alaska are responsible for excreting gold dust (Advameg), in perhaps large enough quantities to form gold nuggets. A prime example is a rod-shaped bacterium named Thiobacillus ferrooxidans that, once discovered, was immediately put to use in the copper, gold and uranium industries, to extract metals from low grade ore in a process they call biomining. Most of these bio-miners are wild crafted, but during the last forty years new strains have been cultured that are resistant to the arsenic, and mercury, currently used in the processing of gold.

For many years, scientists have wondered why metals appear in two forms, solid and dissolved. Gold deposits found in veins of quartz are suspected of being the waste product of mineral eating microbes. In his article "'Gold Bug' Sheds Light on How Some Gold Deposits Formed," Derek Lovely, a University of Massachusetts microbiologist, states "A vast number of bacteria and archaea have the ability to transfer electrons to iron through a reduction process. In other words, they digest one form of a metal and excrete it as another form. This transfer leaves behind deposits of solid metal in unlikely places on Earth, (or maybe even on Mars). What's left behind is often more useful, or more accessible to humans, than the original form of the same substance." (Amherst, 2001) Prior to this theory, science assumed gold was deposited by hot water forced through quartz by volcanic processes, but gold doesn't readily dissolve in water. They have never been sure why this supposed dissolved gold coagulates, and they now suspect it's a result of microbial activity where environmental conditions are optimum for their growth. Interestingly, fault zones, where intense heat and pressure are expected, often have metal deposits in and around them.

Continuing evidence that the Philosopher's Stone is actually comprised of, and/or created by the use of, microbes is probable when you consider the recent discovery that Aerobic microbes are responsible for mineral weathering and erosion. It has been observed that these microbes oxidize minerals resulting in the creation of metal sulfides such as pyrite (iron), galena (lead), chalcopyrite (iron and copper), chalcocite (copper), aresenopyrite (arsenic and iron), and sphalerite (zinc). (Parkes) I believe scientists will eventually find that cinnabar (mercury), a preferred mineral source for transmutations using the Philosopher's Stone, also belongs on this list. In Roman times they collected mineral salts observed in the run-off from their copper mines, and deduced how to more easily extract the copper from the resulting salts. We now know that microbes were responsible for this greenish crystalline form of copper, but until recently it was commonly believed that copper is totally anti-bacterial in nature.

ALCHEMY ASTROLOGY

Astrology is helpful because it tells us when the magnetosphere is being either positively, or negatively, charged, inducing an increase or reduction of electron activity. Planaria Dugesia dorotocethala microbes were sometimes found to heal from injury faster when exposed to low frequency magnetic fields in a laboratory, which can occur naturally under certain astronomical conditions. Some aquatic microbes are magnetotactic bacteria, which detect magnetic fields that help them navigate in the sediment habitat between oxic and anoxic conditions. They produce either magnetite, or greigite, in a magnetic organ in their bodies that allow them align with the Earth's magnetic field. Science uses their remnants to deduce the paleomagnetic record of Earth's evolution. Minerals such as carbonates and dolomites, gypsum, uraninite, apatitie, siderite, iron oxide, manganese oxides, silica and mixed iron-aluminium silicates, are also associated to these lithoautotroph microbes. Therefore, choosing which microbial source to work with, and identifying when they are most active, can be determined by the strength of the predominate astronomical influences present.

Extremophile microbes are also being researched for medicinal uses with very positive results. They were tested for their ability to kill cancer cells, and malaria, according to Diana Northup, a microbiologist and associate professor at the University of New Mexico. (Tyson & Northup) Some microbes even live in areas teaming with radiation from minerals such as uranium, and may be a useful adjunct to radiation therapy.

A newly established scientific project (HMPDACC, 2010) titled the "Human Microbiome Project," (NIH) in association with the Human Genome Project, has published a list of 178 microbes that live in our bodies. Though it has been known since 1864 that some microbes cause disease (BBC, 2010), scientists believe that most of the microbes in our bodies are necessary for our good health. It is theorized that without the influence of these microbes the evolution of human life may have been impossible (Singer, 2007). The HMP is currently initiating studies that will sample the microbiomes of healthy volunteers, and volunteers with specific diseases (Mora, 2010).

Modern scientists, who are exploring the benefits of microbes for health and industry, are having great difficulty transferring their samples to the laboratory without destroying the microbes they wish to study. It is important that the mineral be virgin and that any microbial life contained within be sheltered, and protected from contamination. The alchemists knew that when harvesting these minerals they must be sheltered from sunlight, air, and cosmic rays. They were sealed in clay vessels and were then only opened when the laboratory environment most matched their natural one. Alchemists also knew what kind of environment these microbes needed to thrive. Science is still working on that problem. It is the collection and multiplication of these microbes that are the principal agents of the Philosopher's Stone.

Just as modern metallurgists use the exact alchemical formula for a liquid historically called the Aqua Regia to purify the noble metals (a 1:3 mixture of concentrated nitric and hydrochloric acids), I suggest scientists look to other alchemical processes in their quest to study microbes. The alchemists discovered how to properly harvest, germinate, and cultivate, what they often referred to as the Seed of the Metals (mineral microbes) in their laboratory furnaces. They discovered how to replicate the hot and vaporous natural environments found deep within the earth where microbes thrive, crystals grow and metals coagulate. This information is available and many alchemists are currently experimenting in small laboratories around the world.

It is my conclusion that the Philosopher's Stone houses a very primal microbial spore, and it provides the universal food that can promote its growth and multiplication. Perhaps it is the first microbe that sparked all life on our Earth. It is understandable that the Stone could transform one mineral substance into another, and initiate medicinal cures in the physical body. Mercury and Gold are next to each other on the Periodic table, they have only one electron different: Hg80 (Mercury), Au79 (Gold). Since, a vast number of microbes can transfer electrons to iron, (Amherst, 2001) why not a microbe that voraciously consumes heated mercury (gently extracted from cinnabar) and in doing so an electron is removed from the Mercury to change the remaining molecules to Gold? The Human Microbiome Project has reported that microbes out number the cells in our bodies, ten to one, and propose that they are necessary for maintaining good health. So, why not alchemically explore these new findings? As recent as 1960 the theories of Plate Tectonics, and life on other Planets, were considered to be as unscientific as alchemy. These ideas were shunned, and discredited, or at best were referred to as "radical science," but now we think differently. Let's open our minds and look at alchemy as radical science, and in so doing bring this valuable information from the past, forward into the present, for the greater benefit of humankind.

7. The Adept Practitioner

The art of alchemy has always depended upon the ability of the practitioner to observe extremely subtle matter resources. These natural materials are best found from locations that have remained untouched by human intervention. They must come to know the strongest premium plants to harvest, and learn to identify the correct type of organic, or mineral deposits, to collect. To preserve them, purify their environment, and provide the perfect conditions they need to grow, takes experience.

Every plant has yeasts growing on it that either aid or hinder them. The alchemist must choose the healthiest plants, or the fermentation will sour, and the plant spirit (alcohol) is lost. Like yeasts, each extremophile colony is made up of many different microbes, and viruses. The talent is to be able to isolate the best colony to help create/grow the Philosopher's Stone. The methods vary, and there are formulas and classes available upon which to base your experiments. Each alchemist builds on their knowledge by consulting historic and contemporary alchemical laboratory writings, and repeating these recorded experiments. Remember to keep good laboratory notes so that you can add to this knowledge.

ALCHEMY ASTROLOGY

As you do this work, the deeper you go into alchemy, it is said that your own vibration will rise. Your mind will become calm, your awareness will grow, and your spirit will blossom. Acting upon the matter by way of separation, purification, and cohobation, will gestate the same process of heightening vibrational rhythms within you. Alchemy is a Spiritual journey that is proven to us by our observations in the laboratory, and will lead to the purification, and multiplication, of our own Inner Gold. Astrology can help to guide you on this journey.

"If I have seen farther than others, it is because I was standing on the shoulders of giants."
-- Isaac Newton

CHAPTER TWO – ASTROLOGICAL CHARTING

1. How To Make A Basic Tropical Zodiac Astrology Chart

The wheel of the zodiac is divided into 12 equal parts. Each section is 30 degrees and represents a different constellation of stars. The resulting 360 degree circle is called the ecliptic plane of the planets in orbit around the Sun. Envision a dinner plate with the Sun on the center and the planets traveling counter-clockwise around it. The zodiac constellations are comprised of stars situated outside of, and in line with, the edge of the plate. Other constellations exist above and below the plate, but for now we are most concerned with the ones on this ecliptic plane.

If you draw a line from the Earth and extend it through the Sun to the edge of the plate it will point to the constellation that affects us on a given date. The Sun acts like a giant magnifying glass that focuses the energies emitted by the constellations. Modern science calls this phenomenon "gravitational lensing." Astrologically, planets also focus energies in a similar way when they are 60, 90, 120, and 180, degrees from each other. Astrologers call these the "aspects." The basic aspects are termed: sextiles (60°), squares (90°), trines (120°) and oppositions (180°). Planets can also be in line with each other; these are called conjunctions.

The aspects combine and amplify the energies of celestial bodies. Alchemist Manfred M. Junius wrote, "John H. Nelson, electro-engineer at RCA, has discovered that magnetic storms harmful to health occur when planets stand at angles of 0, 90, and 180 degrees to each other. In connection with measurements of radio interference, he has proved that interference-free fields are produced when planets form angles of 60 and 120 degrees." (Junius, 1985)

To make a chart you will need to use an ephemeris to look up the positions of the planets as associated with the 12 zodiacal divisions. An example is provided on Page 12, for instructional purposes. Next, make a copy of the full-page blank chart in the appendix at the back of this book. You can also download pdf files from the Swedish Ephemeris at: www.astro.com/swisseph/swepha_e.htm.

Begin with the section on the left side of the Wheel just below the horizontal line. There are two ways to label your chart. You can start with the zodiac symbol of the Sun's position on the date you're charting, or, you can begin with your Sun sign as listed in an Ephemeris on your birth date. Continue counter-clockwise and draw the remaining 11 signs in order, one sign per section. This is also the direction the planets travel around the chart. If the Sun is listed in Scorpio, start with Scorpio and end with Libra. The signs in celestial order are: Aries, Taurus, Gemini, Cancer, Leo, Virgo, Libra, Scorpio, Sagittarius, Capricorn, Aquarius, and Pisces.

When you place the planets in a chart according to ephemeris listings you are creating a tropical chart. This type of chart is based on the seasons with the spring equinox marking the beginning of Aries. It is commonly used for charts depicting the influences present when you were born. It is a valid tool for self-realization associated to personal, and group, interactions within our community, and is worthy of investigation. To know yourself is an important part of becoming a good alchemist. Later I will show how to adjust this information to create what is called a sidereal chart. This type of astrology is based on the direct influences of the star constellations. It will add to your individual self-discovery, and is a more valuable tool for work in the laboratory.

I have chosen an interesting date as an Ephemeris example. On the winter solstice, December 21, 2012, the wobble of the Earth's axis comes around to point towards the center of our Milky Way galaxy. Modern Astronomers have observed that an intensely gravitational giant black hole resides there. Our solar system is also aligning with the radiation disk of our Milky Way Galaxy, which occurs approximately every 13,000 years. It is believed that this alignment can be attributed to a gradual increase in earthquakes (HEWS, 2010), volcanoes (Smithsonian Institute, 2010), and unusually intense and unpredictable weather conditions. It is also the last day of an ancient Mayan calendar, which some believe represents either, the "End of Days," or the beginning of the "Golden Age." It is also likely that a sudden increase in sunspot activity, predicted to occur relatively near this same time period, will also be of great concern.

If you use your birth date for practice, start with your Sun sign in the first position. I was also taught to put my Sun sign first when I'm making charts for laboratory work. This makes it easier to compare it to my personal chart, and it also reminds me how, as the alchemist, my energy affects my results. In the book, *The Tao of Physics*, author Fritjof Capra, Ph.D., physicist and systems theorist, writes about how two scientists, who are watching the same nuclear particle experiment, can record different observations. The results of any such delicate process are dependant upon the qualities of the observer. This is also true for alchemy.

ASTROLOGICAL CHARTING

2. Placing The Planets

Using the Ephemeris example provided I have placed the planets in their approximate tropical positions. My tropical Sun sign at the time of my birth is Scorpio, so, as the alchemist preparing the chart, I listed it first.

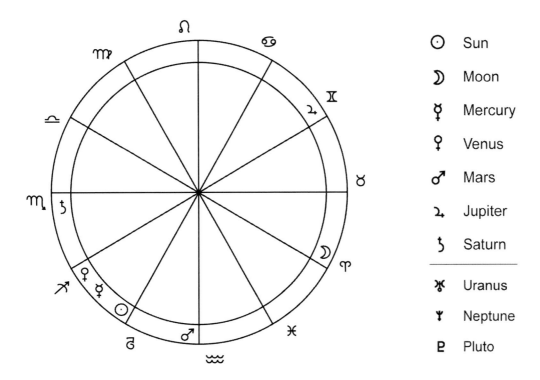

I have not placed the three outer planets in this book – Uranus, Neptune, and Pluto - because they cannot be seen with the naked eye. Ancient alchemists didn't know of their existence, and therefore, these planets were never researched. I invite you to explore their possible influences in the lab. They move very slowly through the cosmos, and I'm sure they add to the energies of the other planets, but I have not found the connection to processes in my herbal lab work, or in the research of other alchemists. I have found they can be helpful as they apply to a personal birth chart though. So, for now, I will proceed based on historical precedence, as applied to alchemy.

DECEMBER 2012 — LONGITUDE

Day	Sid. Time	☉	0 hr ☽	Noon ☽	True ☊	☿	♀	♂	♃	♄	♅	♆	♇
21 F	5 59 52	29♐31 27	8♈28 49	14♈38 43	25♏36.2	14♐0.5	6♐0.2	26♉3.8	8♊57.4	8♏37.1	4♈38.3	0♓48.2	8♑55.9
22 Sa	6 3 49	0♑32 33	20 44 32	26 46 53	25 37.6	15 29.3	7 15.2	26 50.7	8 50.3	8 42.5	4 38.6	0 49.5	8 58.0

ALCHEMY ASTROLOGY

3. The Aspects

I was taught in alchemy school that the aspects are drawn between the zodiac signs that are populated by planets, regardless of the planets precise location within each sign. A sextile spans 3 signs, a square spans 4, a trine spans 5, and an opposition spans 6 (directly opposite). A conjunction is 2 planets in the same sign. The following example chart is done in this manner:

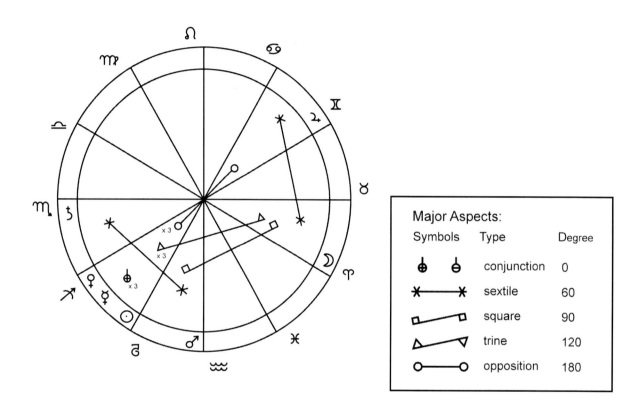

Where there is more than one planet per zodiac sign we will have more than one aspect. The sign Sagittarius has three planets that are in conjunction with each other: Venus is conjunct Mercury, Mercury is conjunct the Sun, and the Sun is conjunct Venus. Mercury, Venus, and the Sun, are all in trine with the Moon, and in opposition to Jupiter. Therefore we count these as three conjunctions, three trines, and three oppositions. I have denoted this by placing "x 3" next to the aspect symbol. If you prefer a more visual representation you can draw three aspect symbols.

Contemporary astrological aspects are only considered when planets are within a specific range of degrees. This range is called an Orb. For instance, oppositions are counted when they span 6 signs, and the planets positions are less than 7° apart. An example of this is Jupiter at 8 Gemini 57.4, and Venus at 6 Sagittarius 0.4, thus they are said to be in opposition at less than a 7 degree orb (actual difference = 2° 57" 0'). Squares, trines, oppositions, and conjunctions, are counted when they are within a 7° orb, sextiles when they are within a 5° orb.

ASTROLOGICAL CHARTING

Later I will present the different influence for each aspect. The aspects in the following chart are calculated using the orb rules.

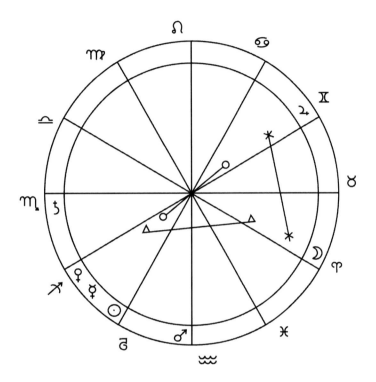

DECEMBER 2012 — LONGITUDE

Day	Sid. Time	☉	0 hr ☽	Noon ☽	True ☊	☿	♀	♂	♃	♄	♅	♆	♇
21 F	5 59 52	29♐31 27	8♈28 49	14♈38 43	25♏36.2	14♐0.5	6♐0.2	26♑3.8	8♊57.4	8♏37.1	4♈38.3	0♓48.2	8♑55.9
22 Sa	6 3 49	0♑32 33	20 44 32	26 46 53	25 37.6	15 29.3	7 15.2	26 50.7	8 50.3	8 42.5	4 38.6	0 49.5	8 58.0

15

ALCHEMY ASTROLOGY

4. How To Make A Sidereal Zodiac Astrology Chart

DECEMBER 2012   LONGITUDE

Day	Sid. Time	☉	0 hr ☽	Noon ☽	True ☊	☿	♀	♂	♃	♄	♅	♆	♇
21 F	5 59 52	29♐31 27	8♈28 49	14♈38 43	25♏36.2	14♐0.5	6♐0.2	26♊3.8	8♊57.4	8♏37.1	4♈38.3	0♓48.2	8♑55.9
22 Sa	6 3 49	0♑32 33	20 44 32	26 46 53	25 37.6	15 29.3	7 15.2	26 50.7	8 50.3	8 42.5	4 38.6	0 49.5	8 58.0

The second column of our Ephemeris example is titled *Sid. Time*, which is an abbreviation for Sidereal Time - sidereal means "of the stars." It is a calculated adjustment that realigns the zodiac chart with the actual position of the star constellations. Add the sidereal time number to the positions listed for the planets, then subtract one star sign by moving clockwise around the zodiacal wheel to the previous sign. (Planets move counter-clockwise.)

Notice that the sidereal time, and the Sun, aren't listed as decimals; they are in degrees, minutes and seconds. The minute and second calculations for the planets are converted to decimals, probably to save space. For the purpose of creating our basic chart it's not necessary to compute the seconds, but it is important when using the aspect orb rules, and for making accurate notes for future reference.

On the 21st of December the sidereal time is 5 degrees, 59 minutes, and 52 seconds. Let's add this number to the positions of Jupiter and Mars. You can convert the decimals by multiplying the number after the decimal point times six. When you add seconds to seconds, any number bigger than sixty totals one minute, which is carried to the minute column. Any number that totals more than sixty minutes totals one degree, which is carried to the degree column. Any number that totals more than 30 degrees puts the planet in the next sign.

Examples:

Jupiter, 8 57 .4 (4 x 6 = 24 seconds)

     8 57 24 in Gemini

+   5 59 52 (Sid. Time for 12/21/2012)

     14 57 16 in Gemini - subtract one sign and your result is 14 57 16 in Taurus.

When the degrees total more than 30, subtract 30, and leave the planet in its originally listed tropical sign.

Mars, 26 3.8 (8 x 6 = 48 seconds)
    26 03 48 in Capricorn
+   5 59 52
    32 03 40
−   30 00 00
    2 03 40 in Aquarius - subtract one sign and your result is 2 03 40 in Capricorn.

From these two examples we find that Jupiter moves to a different sign, which changes the aspects, while Mars stays in the same sign it was in tropically. The following image is the sidereal rendering using the same ephemeris example as before. Normally I would put my sidereal Sun sign, Libra, in the first position. In the interest of making it easier to compare the tropical example to the sidereal example, I have left Scorpio first. (Some people are born under the same Sun sign for both tropical and sidereal charting, only the degree changes.)

In 'Appendix A' there is a blank chart that can be printed, or photo-copied, onto a clear plastic sheet. You can purchase the transparent sheets that are appropriate for your printer/copier at an office supply store. Place it over the tropical chart you made in Chapter Five, and put Libra over Scorpio. Essentially you are subtracting one sign first. To add the sidereal time turn the transparency clockwise the number of degrees listed in the sidereal time column. Trace the planetary positions with an erasable marker and you have your sidereal chart. Remember, this could change the aspects.

ALCHEMY ASTROLOGY

The following image is the Sidereal rendering without using orb rules for the aspects. This gives us a broad feel of all the possible energetic combinations.

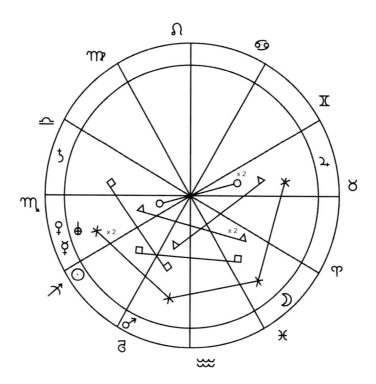

The next image is the Sidereal rendering using the orb rules for the aspects. This information highlights aspects that have the greatest focus of energy.

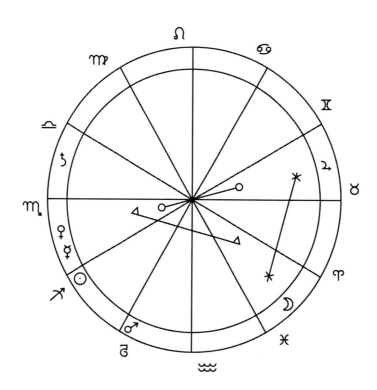

CHAPTER THREE – THE FIVE LAWS

The Five Laws

1. Polarity of the Planets
2. Influences of the Zodiac
3. Duration of the Influences
4. The Four Elements
5. The Aspects

Using the Five Laws we can initially represent the zodiac chart as two numbers: total positive influences, and total negative influences. The positive number represents the amount of active principles present, and the negative number represents the passive principles. It is a good day to begin the work when the total count has more positives than negatives. After determining these numbers we can then contemplate the particular meanings inherent in the chart. Through investigation, and careful documentation, we can deduce additional information that can be helpful for future experiments.

In the first part of this Chapter I will use the Sidereal example without the orb rules. Then I will show how the orb rules change the count, thus focusing our attention on only the strongest energies available. In Chapter 9, Evaluating a Chart, we will look at possible interpretations of the information gathered from the Five Laws.

ALCHEMY ASTROLOGY

1. Law One - Polarity Of The Planets

Each planet, and every zodiac sign, are either positive or negative. The following image is a list of these polar influences:

Polarity of the Planets & the Zodiac

Planets		Zodiac		
+	☉ Sun	+	♈	Aries
-	☽ Moon	-	♉	Taurus
+/-	☿ Mercury	+	♊	Gemini
+	♀ Venus	-	♋	Cancer
-	♂ Mars	+	♌	Leo
+	♃ Jupiter	-	♍	Virgo
-	♄ Saturn	+	♎	Libra
		-	♏	Scorpio
		+	♐	Sagittarius
		-	♑	Capricorn
		+	♒	Aquarius
		-	♓	Pisces

Of the seven major planets, three are positive, and three are negative, with the planet Mercury as variable. Therefore we either have four positives and three negatives, or three positives and four negatives with regards to the first law. Mercury takes on the polarity of the closest planet present in the same sign. If Mercury is alone in a sign it takes on the polarity of the sign.

In the Sidereal chart example (not using aspect orb rules) Mercury is positive because positive Venus is its closest influence. The following example depicts how the polarity of the planets are written when Mercury is positive:

THE FIVE LAWS

The Five Laws
1. Polarity of the Planets
2. Influences of the Zodiac
3. Duration of the Influences
4. The Four Elements
5. The Aspects

	+	-
1.	4	3
2.		
3.		
4.		
5.		

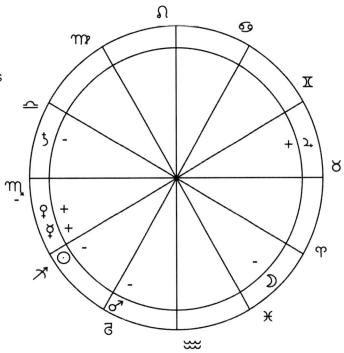

ALCHEMY ASTROLOGY

2. Law Two - Influences Of The Zodiac

For the second law we consider the polar influences of the zodiac. First we count how many positive and negative zodiac signs have planets in them. Later we contemplate the general descriptive influences of these signs. In our Sidereal chart example we find two positive signs, and four negative signs, that have planets in them.

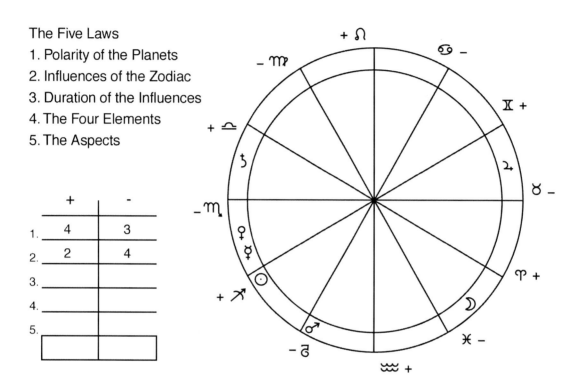

The solar system spins and vibrates at a frequency that creates a pattern of energy sectioned into twelve angular slices. This creates a circular wave of energy with a positive to negative alternating nature. Each piece is influenced by the gravitational lensing of energies from the stars that are located behind the Sun; both, the positive/giving signs, and the negative/receiving signs.

Many examples of this frequency effect, using sand on a vibrating piece of flat metal, can be viewed online. One in particular shows a circle pattern dividing into six slices, as the frequency was increased it divided into 12 dual parts. (Mountain Waves Healing Arts, 2009)

THE FIVE LAWS

3. Law Three - Duration Of The Influences

The third law gives us a feeling for time and continuity regarding the influences present. Duration is tallied the same way that the second law (Influences of the Zodiac) is counted. Fixed timing indicates stable long-term accumulative influences. When energies are mutable they are medium in length of time, and ripe for changes that are versatile and adaptable. Cardinal timing is short in length, and good for transitions, and initiating activities. Each of the zodiac signs is associated with one of the three durations.

Fixed – Taurus, Leo, Scorpio, Aquarius.
Mutable – Gemini, Virgo, Sagittarius, Pisces.
Cardinal – Aries, Cancer, Libra, Capricorn.

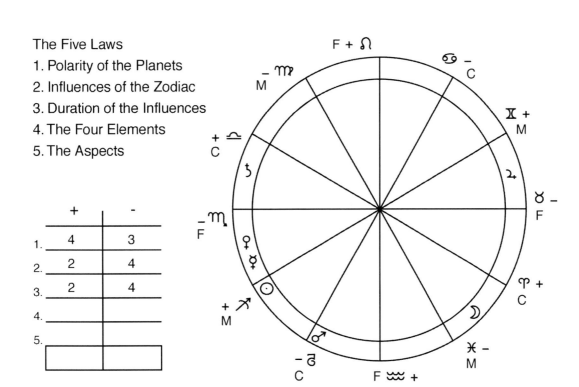

ALCHEMY ASTROLOGY

4. Law Four – The Four Elements

The fourth law pertains to the four elements: fire, earth, air, and water. The elements are tallied the same way that the second law (Influences of the Zodiac) is counted. Fire and air are considered to be positive/active, aggressive and prone to give forth, whereas water and earth are negative/receptive and passive.

Element	Alchemy	Science
Fire	radiant and etheric	strong nuclear force
Earth	solid and concrete	gravitational force
Air	gaseous and volatile	electromagnetic force
Water	liquid and fluid	weak nuclear force

Alchemy says that all four elements can unite in one element, and that each element can be found within each element. For example, you can distil water to extract the fire of water, the earth of water, the air of water, and the water of water. Also, hidden within the four elements, but not one of the four, is found the fifth – the quintessence. Astrology can give us clues for finding and purifying this quintessence.

The elements are further defined by the properties dry, cold, moist and hot. Fire combines with earth to equal dry, earth with water to equal cold, water with air to equal moist, and air with fire to equal hot. This information is valuable when we make our overall evaluation of the chart. The following graphic visually depicts how the elements in a chart can combine to create different influences:

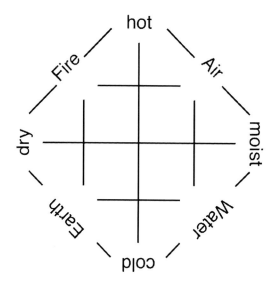

Each of the zodiac signs is associated with one of the four elements.

Element	Zodiac Sign
Fire	Aries, Leo, Sagittarius
Earth	Taurus, Virgo, Capricorn
Air	Gemini, Libra, Aquarius
Water	Cancer, Scorpio, Pisces

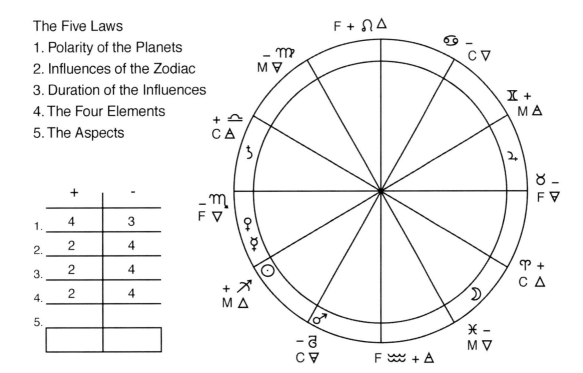

The Five Laws
1. Polarity of the Planets
2. Influences of the Zodiac
3. Duration of the Influences
4. The Four Elements
5. The Aspects

	+	−
1.	4	3
2.	2	4
3.	2	4
4.	2	4
5.		

If you are working with an herb governed by the Sun, and the Sun is in a fire sign you may want to start slowly, and incrementally increase your heat source. If you're calcining under such influence, take extra care that you don't melt the organic metals out of the salts. When the crucible is too hot, a mirror like coating will appear on the bottom of the dish, and the organic metals will be lost as the Salts diminish in quality. Similarly, if the Sun were in Leo, you may find that fermentation is difficult, because it has to much fire, or to much cosmic heat.

ALCHEMY ASTROLOGY

5. Law Five - The Aspects

The fifth law shows us how all the influences are inter-twined, softened, or intensified, by their relationship to one another. Sextiles and trines are counted as positive/active influences, squares and oppositions are counted as negative/passive influences. Conjunctions are either positive or negative depending on the polarity of the planets in conjunction. Two positive planets in conjunction (in the same sign) form a positive aspect. Two negative planets in conjunction form a negative aspect. When one positive planet and one negative planet are in conjunction the polarity of the zodiac sign decides the polarity of the conjunctive aspect. In the interest of teaching the basics, and using only information available to the ancient alchemists, I have not included the minor aspects used by contemporary astrologers.

Major Aspects:			
Polarity	Symbols	Type	Degree
+ / −	⊕ ⊖	conjunction	0
+	✳—✳	sextile	60
−	▫—▫	square	90
+	△—▽	trine	120
−	○—○	opposition	180

The following image shows the placement of the aspects from our sample date, 12-21-2012, just as the ancient alchemists would have used them (without the orb rules):

THE FIVE LAWS

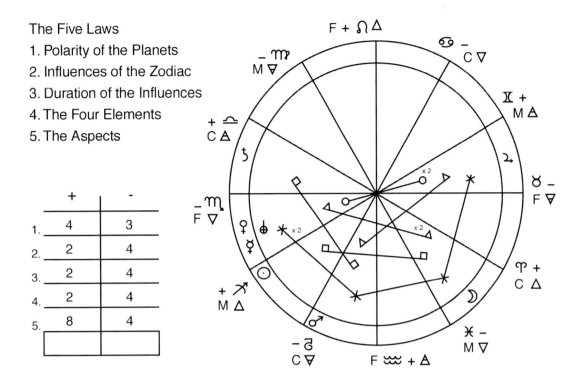

6. The Total Count

Finally, in the following image, we total all the positive (active) numbers (18), and then we total all the negative (passive) numbers (19), to get our final count.

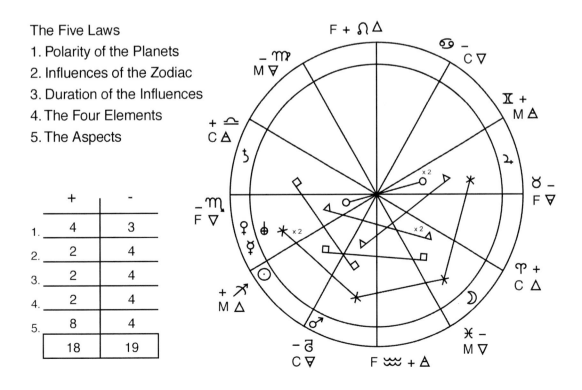

27

ALCHEMY ASTROLOGY

Conclusion

We now see that on Saturday, December 21, 2012, in our sidereal chart there are more negative (-19, +18), or passive, influences. Without careful consideration of the overall picture I would not begin a laboratory project on this date. In a case like this, I would consider the aspects using the orb rules to see if it changes the count. It can also give me a more precise understanding of where strengths and weaknesses may be indicated in relation to my lab project. Unfortunately in the following chart, using the aspect orb rules, the negatives even farther out-weigh the positives. One change that occurs is that Mercury becomes negative in this case because the orb around positive Venus is more than 7°. Mercury now takes on the polarity of the negative sign of Scorpio, instead of the polarity of the nearest planet in the same sign. The other major difference in the count is a reduction in the total number of aspects when we eliminate those that are outside the limits of the orb rules. This reduces the number of positive aspects in our example from 8 to 2, and reduces the negative result from 4 to 1.

The following image is the sidereal chart using aspect orb rules - Total count: 11+ 17- :

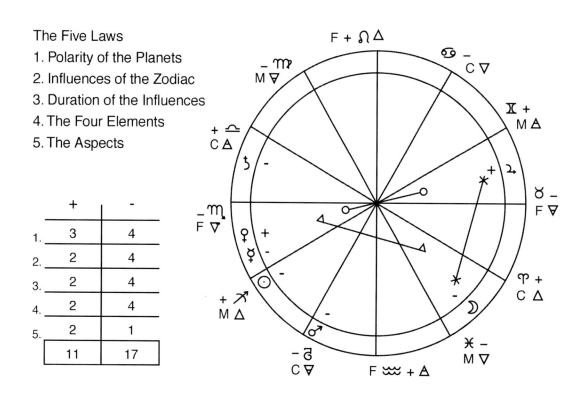

CHAPTER FOUR - INFLUENCES

1. General Influences

With your chart in hand, look at the zodiac signs that have planets in them. Start by contemplating the general influence that each sign emanates, the duration of those influences, and the associated elements. The following tables list these influences as single word concepts.

Symbol	Polarity	Zodiac	Influence	Duration	Element & Symbol
♈	+	Aries	Aggressive	Cardinal	Fire △
♉	−	Taurus	Decisive	Fixed	Earth ▽
♊	+	Gemini	Diffusive	Mutable	Air △
♋	−	Cancer	Tenacious	Cardinal	Water ▽
♌	+	Leo	Organizing	Fixed	Fire △
♍	−	Virgo	Analytical	Mutable	Earth ▽
♎	+	Libra	Uniting	Cardinal	Air △
♏	−	Scorpio	Solidifying	Fixed	Water ▽
♐	+	Sagittarius	Inspirational	Mutable	Fire △
♑	−	Capricorn	Retentive	Cardinal	Earth ▽
♒	+	Aquarius	Concentrative	Fixed	Air △
♓	−	Pisces	Relaxitive	Mutable	Water ▽

Duration of time; Cardinal = short, Fixed = long, Mutable = variable

2. Influences Of The Planets

We then add the general influences of the planets to our previously contemplated information. Then when we look at the associations between the planets, and the energy flows inherent in the aspects, we can find the path of highest energies available for our experiments.

Planetary Influences			
Polarity:	Symbol:	Planet:	Influence:
+	☉	Sun	Persistant
-	☽	Moon	Change
+/-	☿	Mercury	Mental
+	♀	Venus	Pleasant
-	♂	Mars	Impulsive
+	♃	Jupiter	Fortunate
-	♄	Saturn	Disruptive

Length of planetary orbits around the Sun, in Earth days, gives us a clue to the speed with which the influences impart their effect:

Planet	Orbit in Earth days	Orbit in years
Moon	28	-
Mercury	88	-
Venus	224.5	0.75
Mars	664	1.8
Jupiter	4,380	12
Saturn	10,220 – 10,950	28-30

Note: The planet Venus rotates on its' axis in the opposite direction of the other major planets.

Jupiter and Saturn radiate about twice as much energy that they receive from the Sun. They are classified as gas giants, and they don't have any solid matter at their core. (Anil Bhardwaj, 2005)

Planet	X-ray Emissions
Jupiter	0.3-2 GW
Saturn	0.1-0.4 GW
Earth	10 - 40 MW

INFLUENCES

3. Influences Of The Major Aspects

Polarity	Symbols	Type	Degree	Influence
+ / -	⊕ ⊖	conjunction	0	Emphasis
+	✳—✳	sextile	60	Opportunity
-	▫—▫	square	90	Challenge
+	△—▽	trine	120	Flow
-	○—○	opposition	180	Awareness

Note: The active, and intense, energies of the Sun, Mercury, and Saturn (also Pluto), in trine, conjunction, and opposition, are best for the mineral work, where as, the Moon and Venus are better allies in the plant work.

4. Influences & Lab Processes

The following convenient chart associates each zodiac sign with its' related lab process when the Moon is present in that sign. This gives us further indication of the value of a chart by determining which stage of the work we may choose to begin on a given day.

Zodiac	Polarity	Influence	Duration	Element	Lab process - Moon
Aries	+	aggressive	Cardinal	Fire	calcination
Taurus	-	decisive	Fixed	Earth	congelation
Gemini	+	diffusive	Mutable	Air	fixation
Cancer	-	tenacious	Cardinal	Water	dissolution
Leo	+	organizing	Fixed	Fire	digestion
Virgo	-	analytical	Mutable	Earth	distillation
Libra	+	uniting	Cardinal	Air	sublimation
Scorpio	-	solidifying	Fixed	Water	separation
Sagittarius	+	inspirational	Mutable	Fire	incineration
Capricorn	-	retentive	Cardinal	Earth	fermentation
Aquarius	+	concentrative	Fixed	Air	multiplication
Pisces	-	relaxative	Mutable	Water	projection

ALCHEMY ASTROLOGY

This table supplies a brief description of the Lab Processes.

Zodiac Sign	Lab Process	Description
Aries	Calcination	To reduce the ash to a fine whitened powder so that it becomes spongeous and receptive.
Taurus	Congelation	To allow matter in a liquid state to gel and collect together of its' own volition.
Gemini	Fixation	Adding a liquid to a solid so they bind and become one, often resulting in a waxy state.
Cancer	Dissolution	To dissolve a solid into a liquid, such as adding the salts into water.
Leo	Digestion	To allow one substance to merge with another, as in macerating plant matter in water or alcohol, or by cohabiting the alcohol and oil with the salts. Gentle heat is often called for in this process.
Virgo	Distillation	To vaporize a wet substance so that it can be condensed to a liquid.
Libra	Sublimation	The application of heat to vaporized a solid, that can then be collected in a receiving vessel similar to distillation.
Scorpio	Separation	To remove the subtle from the gross, as in the oil from the leaf, or the salt from the body. Filtration through filter paper, or, gravitational separation with the use of a separatory flask.
Sagittarius	Incineration	To burn with fire - impurities are released in the smoke.
Capricorn	Fermentation	To allow the natural yeasts, or to add yeast, to putrefy matter so that it becomes alcohol.
Aquarius	Multiplication	Strengthening by purification.
Pisces	Projection	Adding a small amount of a dry substance to a large volume of liquid so that it dissolves and penetrates the liquid, it is sometimes evident by a change in color.

INFLUENCES

5. Influence With The Time Of Day

DECEMBER 2012					LONGITUDE								
Day	Sid. Time	☉	0 hr ☽	Noon ☽	True ☊	☿	♀	♂	♃	♄	♅	♆	♇
21 F	5 59 52	29♐31 27	8♈28 49	14♈38 43	25♏36.2	14♐0.5	6♐0.2	26♊3.8	8♊57 4	8♏37.1	4♈38.3	0♓48.2	8♑55.9
22 Sa	6 3 49	0♑32 33	20 44 32	26 46 53	25 37.6	15 29.3	7 15.2	26 50.7	8 50.3	8 42.5	4 38.6	0 49.5	8 58.0

The time of day is also an influence when choosing a start time. Listings in the ephemeris are calculated from 0° longitude at midnight. If you live west of 0°, subtract 4 minutes for each degree of longitude. If you are east, add 4 minutes per degree. You can convert the minutes to decimals by dividing the number by six, and moving the decimal point left one digit. To convert the decimal number, right of the decimal point, back to minutes, multiply times six.

Here's an example:

Chicago is located at: 87° 39" 0' W

87.65 x 4 (minutes) = 350.6 minutes

350.6 divided by 60 = 5.843

or, approximately 5 hours and 50 minutes.

Subtract this time from midnight, because this location is West of 0° longitude, and you get 6:10 PM (or 18:10 hours).

This means that the influences of the planets as they are listed at midnight, on 0° longitude, are happening at 6:10 PM in Chicago, the evening before.

There are several online resources for looking up latitude and longitude of the city closest to your location. From a search engine type in: "(your closest city), latitude, longitude." Atlas style road maps sometimes list latitude and longitude.

Timing of the lab work can be further decided by additionally considering the hourly planetary influences. There are several systems of daily influences. The following system was taught at Paracelsus Research College. I have listed the information on a twenty-four hour clock in the reproduction of this chart. A color copy is available at the Alchemy Astrology website. These times are also useful for the harvesting of herbs by association to their governing planet.

ALCHEMY ASTROLOGY

Daily Start Times

	Sun ☉	Mon ☽	Tue ♂	Wed ☿	Thu ♃	Fri ♀	Sat ♄
00:00–03:25	♂	☿	♃	♀	♄	☉	☽
03:25–06:51	☉	☽	♂	☿	♃	♀	♄
06:51–10:17	♀	♄	☉	☽	♂	☿	♃
10:17–13:42	☿	♃	♀	♄	☉	☽	♂
13:42–17:08	☽	♂	☿	♃	♀	♄	☉
17:08–20:34	♄	☉	☽	♂	☿	♃	♀
20:34–24:00	♃	♀	♄	☉	☽	♂	☿

Day	Planet	Color
Sunday	Sun	Yellow
Monday	Moon	Violet
Tuesday	Mars	Red
Wednesday	Mercury	Orange
Thursday	Jupiter	Blue
Friday	Venus	Green
Saturday	Saturn	Black

6. Influence Of The Moon

Modern astrology says, what is started at the new Moon finds fruition on the full Moon. When the Moon opposes the Sun we have a full Moon, and if the Moon is in conjunct with the Sun we have a new Moon. This information is also detailed in the ephemeris; day, hour, minute, and location, for all four phases, are located at the bottom of the page, in the "Phases and Eclipses" section, or online (Fourmilab, 2010). Tides are higher at the new and full Moons because the gravitational influences of the Sun and Moon are in line. In addition, when the Moon is on one of its two nodes, meaning it is level with the solar planetary plane, or ecliptic, the tides are more pronounced. The Moon orbits the Earth at an angle to the orbit of the Earth around the Sun, and it descends and ascends through the ecliptic, therefore the Moon on its' node happens twice a month. Nodes indicate an active positive time especially when they line up with the Sun, which happens about every five and a half months. When the new Moon is on a node we have an eclipse of the Sun. Apolunar (far) and epilunar (near) positions are also considered to be a factor, and are referred to as the apogee and perigee (Fourmilab, 2010). These positions occur because the Moon has an oval shaped orbit. This non-circular orbital path is created by the Earth and Moon revolving around each other; they are considered to be more like a dual planet phenomenon than a planet and its' Moon (Hollocher, 2007). If the full Moon is at perigee, and on a node, we have a total eclipse of the Sun. (Bulgerin, 2002) This can be a very powerful time for laboratory work, some suggest to avoid these times, I suggest that you proceed with caution.

Another good example of lunar influences, which can be considered for use in the laboratory, is that fish are more active during full and new Moons, especially at moonrise and moonset. Some anglers have observed that when the Moon is at its' highest, or lowest point below the visible horizon, the activities of fish increase as well. These positive midpoints are determined as halfway between moonrise and moonset, and an online calculator is available on the Naval Oceanography Portal (USNO, 2010). Fishing calendars are also available online. (Lake-Link, 2010)

The phases of the Moon are considered during planting. Generally speaking, annuals that produce above ground should be planted between New Moon to full moon, and biennials, perennials, bulbs and roots should be planted from full Moon to new moon. Specific times for sowing, and harvesting, can be determined by associating the Moons' position to the element of the zodiac sign in which the Moon is currently residing. Root plants are best processed in the earth signs, leafy plants in the water signs, flowery plants in the air signs, and plants grown for seed in the fire signs. The only exception is that biodynamic farmers, who regularly make use of Moon phase information, do not generally consider the Moon is at its perigee, or closest point, as a good time for planting because the energy is to intense for young sprouts. (Lunarium, 2010)

7. Other Influences

Alchemical processes are very sensitive to energies in the room around your laboratory. Take into consideration the temperature, humidity, and barometric pressure. Although it is good to avoid direct sunlight shining in on your laboratory, another concern is the electric lighting you are using (Magda Havas, 2009). Compact fluorescent bulbs have been shown to raise the blood sugar in diabetics, which is not a good influence, and I would be cautious about using them, nor any bulb other than perhaps a soft white incandescent, or colored bulb. For instance, red bulbs influence chicken's to lay more eggs (U of C), it may be a fun experiment to discover if different light sources influence fermentation. Light from an oil lamp/lantern, or candle, would be best (always have a fire extinguisher nearby).

Plants are found to be responsive to activities in their environments (Gill, 2010), to threats (Goudarzi, 2007), and very likely the emotions of humans. I would speculate that microbes are also sensitive to our frequencies, our physical, and emotional vibration. Negative feelings may sabotage our experiments by projecting a deactivating influence onto the microbes.

CHAPTER FIVE – INTERPRETING THE CHART

Each lab project varies because the astrology is never the same, but when you meditate and contemplate the energies of the zodiac represented in the chart you'll come to better understand their significance. Laboratory notes, and time logs for each step are very important so that you can compare one experiment with another. Intuition regarding each process is strengthened and successful operations become the norm. If I had not kept thorough records, I would not be able to repeat a beneficial formulation with any consistency. In fact, I couldn't have written this book with the kind of conviction it deserves.

When you start using astrology regularly, you can compare the influences in the chart to what you perceive is happening in the world. Listen for reports of illness from friends, or in news reports, which are presently active in the general population and see if you can make any correlations. It can be helpful to list current events and trends in the general society, and in your own life, for later comparison to the chart. Similarly, notice any extreme or unusual occurrences that can be associated to what is happening among the planets: geomagnetic and solar storms, solar x-ray flux (NOAA, 2010), sunspots (SOHO, 2010), weather, seismic events, and electronic disruptions, are some examples. This will help you to understand the specific influences of the cosmos. In Chapter Seven you'll find tables that are useful for comparing your chart with the cosmic energies related to the herb or mineral you're working on, and the area of the body you wish to affect.

Let's look at some simple interpretations of the sidereal example chart, the one that uses orb rules for the aspects. I start by forming sentences with the listed influences of each planet, zodiac sign, polarity, duration, element and aspect – these words appear as bold texts. Tables listing other associations to herbs, metals, and the body, are provided in Chapter Seven – Relating To Matter.

In the next section of this book I will show you how the stars themselves complete the process in our decision to begin the work.

ALCHEMY ASTROLOGY

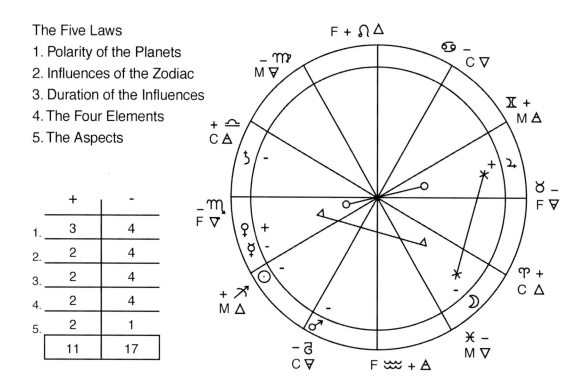

The Five Laws
1. Polarity of the Planets
2. Influences of the Zodiac
3. Duration of the Influences
4. The Four Elements
5. The Aspects

	+	−
1.	3	4
2.	2	4
3.	2	4
4.	2	4
5.	2	1
	11	17

1. The *persistent* Sun is in the *fiery*, *mutable*, and *inspirational* sign of Sagittarius. Unfortunately it is not in aspect to any other planets and so its' energy is less activate, and not available to counter any negative aspects present. The Sun is the most powerful influence on our planet Earth, and good aspects with the other planets are usually worth waiting for.

2. The *changing* Moon resides in the *watery*, *relaxing*, *mutable* sign of Pisces, indicating the lab process of *projection*. This opportunity is strengthened by the Moon's fortunate connection to the *fixed*, positive energy of Jupiter, and energy *flows* with a *pleasant* aspect to positive Venus. Working with an herb governed by the Moon might illuminate complimentary influences. The Moon is in its quarter phase, increasing its' light on the path towards a full Moon - this is a waxing gibbous Moon; sometimes referred to as the Alchemist's Moon. (Note: If the Moon was in Virgo, it would be called a waning, crescent Moon, as its' light decreases on its' path towards new Moon.) This could be a good time to plant or harvest herbs whose leaves are indicated as a remedy. In reference to body and zodiac/planet associations these influences would be beneficial for anyone suffering from respiratory or circulatory afflictions, and would support brain function during recovery.

3. Fortunate positive Jupiter is fixed in earthy, decisive, Taurus, bringing awareness by its pleasant connection to positive Venus who is fixed in watery, solidifying, Scorpio, with the mental influence of Mercury alone at its side. This could be a good time to work with the root of an herb like dandelion, which is governed by Jupiter, and is residing in an earth sign. Considering the earth and water influences, and the

Sun in a fire sign, the processes of incineration, and calcination of salts from leaching, could be the best choice if you have to work on this date.

4. Disruptive Saturn is united to airy, cardinal, Libra, but is alone and captured in this energy. It is good that Saturn is not involved by aspect to the other planets. Saturn can be trouble.

5. Impulsive Mars, also alone in its influence, is just starting its retentive journey through earthy Capricorn. Mars, like Saturn, is fortunately not involved by aspect to the other planets. It is in an Earth sign, which may further reduce the impulsive intensity Mars can sometimes generate. Saturn, and Mars, can add some challenges to our experiments, but, if you are cautious, they can bring you face-to-face with new discoveries too.

Major Aspects:

Polarity	Symbols	Type	Degree	Influence
+ / −	⊕ ⊖	conjunction	0	Emphasis
+	✶—✶	sextile	60	Opportunity
−	▭—▭	square	90	Challenge
+	△—▽	trine	120	Flow
−	○—○	opposition	180	Awareness

Conclusion

According to our totals of positive and negative influences divulged in our example chart, December 21st, 2012, is generally not a good day to 'start' the work, but you may pinpoint some progress continuing projects already begun.

If you have time restraints that implore you to proceed with your project despite less than favorable conditions, you may find solace by comparing the lab chart to your birth chart. The personal astrological energy you are born with can sometimes bring balance to any lack of positive influences present.

CHAPTER SIX – THE STARS

1. The Stars of the Ancients'

I begin with the stars the ancients knew, the ones that can be seen without the benefit of that "modern" invention we call a telescope.

When a star, or planet, appears to be at its' highest level above the horizon, it is at its "culmination point." When stars and planets co-culminate along the same vertical longitude line, at the same moment, they are said to be aligned in "paran" with each other. This influential co-culmination occurs on the opposite longitude line as well. Stars and planets are also in paran when they rise or set at the same time. There are other parans that can be calculated by a computer program but we will focus on the co-rise, co-set, and co-culminating parans, because they can be seen with the naked eye. The opposite co-culmination, which is not visible in the night sky, was easily deduced by ancient peoples just as they did when determining which zodiac sign the Sun was in. Stars visible in the night sky are not the stars that influence the Sun. The stars that represent the sign where the Sun resides are behind the Sun during the daytime. For the purpose of teaching the basics I will begin with the four parans the ancient alchemists may have used.

The tropical paran charts in this handbook are copied from Bernadette Brady's book *Star and Planet Combinations*. She has graciously given me permission to include three latitudes – 40˚, 45˚, and 50˚, north. Other latitudes, both northern and southern, are listed in her publication. Also, she has some very useful information to add to your own personal birth chart, sometimes called a natal chart.

When we convert tropical locations to sidereal positions, we are actually only moving the Wheel of the Zodiac. If you are making a sidereal chart, use the tropical locations from the Ephemeris to find where the stars are, and simply draw them near the planet. Even though we are using the tropical Ephemeris example, the stars are still matched to the planets in the sidereal chart.

Copy the blank chart found in 'Appendix A'. Place the planets from our Ephemeris example in the inner ring. Starting in the left column of the paran chart, look for a match that is within 2˚ either direction of a planet (a 2˚ orb) tropically listed in the ephemeris. Add this star to the outer ring of your chart. Then, in the same left column, look for a star position exactly opposite of the sign at the same degree. These are the co-culminating influences of the stars that are in combination with the planets. The next step is to match the position of the planet with the list of co-rising and co-setting stars nearest your latitude. The following chart shows the stars in position with the planets.

ALCHEMY ASTROLOGY

Sidereal chart showing planets and stars

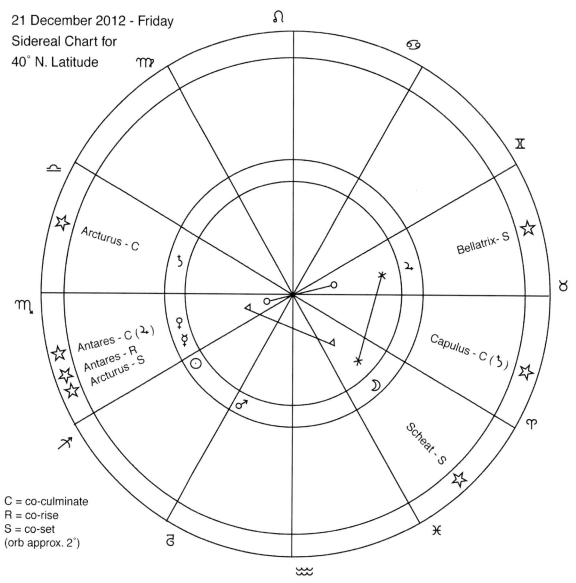

21 December 2012 - Friday
Sidereal Chart for
40° N. Latitude

C = co-culminate
R = co-rise
S = co-set
(orb approx. 2°)

© 2008
Tim Wilkerson

The Five Laws
1. Polarity of the Planets (+ / -)
2. Influences of the Zodiac
3. Duration of the Influences (of the Zodiac)
4. The Four Elements (fire, earth, air, water)
5. The Aspects:

trine + (orb 7°)
sextile + (orb 5°)
opposition - (orb 7°)
square - (orb 7°)
conjunction +/- (orb 7°)

		+	-
Sun	+	1. 3	4
Moon	-	2. 2	4
Mercury	+/-	3. 2	4
Venus	+	4. 2	4
Mars	-	5. 2	1
Jupiter	+		
Saturn	-	11	17

Moon in:
Aries	+	calcination
Taurus	-	congelation
Gemini	+	fixation
Cancer	-	dissolution
Leo	+	digestion
Virgo	-	distillation
Libra	+	sublimation
Scorpio	-	separation
Sagittarius	+	incineration
Capricorn	-	fermentation
Aquarius	+	multiplication
Pisces	-	projection

THE STARS

In 'Appendix A' you'll find a transparency chart that can be machine copied on a printable clear sheet, and placed over your tropical chart. Line-up the transparency so that Libra is directly over Scorpio (this subtracts one zodiac sign), and turn it clockwise to add the number of degrees listed under "Sid. Time" in the Ephemeris. You can then view the sidereal locations of the planets and stars, on the transparency, and trace them with an erasable marker. (Note: the aspects, and the positive / negative count may change.)

You can print another transparency for your birth chart. Use an erasable marker and trace your birth planets. You can then lay this chart over your start-time chart to compare the presently available energies, with the energies you are born with. This can reveal a personal advantage that can counter any negative aspects that would normally prevent a smooth application of alchemical processes.

ALCHEMY ASTROLOGY

2. Star Influences List

Here is a list of star influences deduced from centuries of observations, as they appear in Bernadette Brady's book, *Star and Planet Combinations*. I recommend her book as a companion to this one. Tables listing other associations to herbs, metals, and the body, are provided in Chapter Seven – Relating To Matter.

Stars	Influences *
Achernar	Rapid changes
Acubens	The love of life
Alkes	To carry a precious thing
Alphard	Obsession
Alpheratz	Speed and freedom
Antares	Obsessive passion
Arcturus	A path-finder
Bellatrix	Success with stress
Capulus Nebula	Penetrating action
Deneb Algedi	The law giver
El Nath	Focus attack
Facies Cluster	Penetrating vision
Fomalhaut	Idealism
Menkar	Open to the collective
Procyon	Things in transition
Ras Alhague	The healer
Regulus	Success but vengeance
Sadalsuud	A natural rapport, luck
Scheat	Searching for truth
Sirius	The quest for immortality
Spica	To be gifted
Thuban	To hoard or guard
Vega	Enchantment
Vindemiatrix	To collect and gather

3. Interpreting A Chart Including The Fixed Stars

We can now evaluate the chart again. Keep in mind, when a star is in paran with a planet, their energies are amplified, and general influences are more specifically augmented.

1. The persistent Sun is in the fiery, mutable, and inspirational sign of Sagittarius. It is not in aspect to any other planets. No stars in paran.

2. The changing Moon resides in the watery, relaxing, sign of Pisces, indicating the lab process of projection. This opportunity is strengthened by the Moon's fortunate connection to Jupiter and Energy flows with its pleasant aspect to Venus. The Moon, and therefore the aspect to the Moon, are further charged by the co-rising star Scheat whose influences are associated with motor skills, muscle memory, and searching for the truth. Scheat is an M class red giant star. This type of star is older, and has a lower temperature, and therefore is considered to be 'metal rich', as evidenced by spectroscopic analysis.

3. Fortunate Jupiter is fixed in earthy, decisive Taurus, bringing awareness by its pleasant connection to Venus who is fixed in watery, solidifying Scorpio, with the mental influence of Mercury alone at its' side. Bellatrix co-culminates with Jupiter, relieving heavy stress. This is a good aspect for working with herbs that affect the shoulders, neck and lungs. Bellatrix is a hotter B class blue giant. Antares is also in paran with Jupiter increasing Energy in the immune system in its connection to the kidneys (Libra). Antares is a metal rich M class red supergiant with a Blue companion.

4. Disruptive Saturn is united to airy, cardinal Libra, but is alone in this energy. Acturus and Capulus in Aries, bring help in the form of penetrating, stable, and driven focus on your path. The cartilage of the knees, and testosterone, are of interest here. Arcturus is a K class orange giant. Capulus is a Nebulae.

5. Impulsive Mars, also alone in its' influence, is just starting its' retentive journey through earthy Capricorn. No stars in paran.

Make and use your own evaluation, and upon reflection of the known astrological factors, you can begin to see the focused energies available for your direction in the lab.

ALCHEMY ASTROLOGY

Conclusion

According to our original totals of positive and negative influences divulged in our example chart, December 21st, 2012, was generally not a good day to 'start' the work, but when we consider the specific stars within the zodiac signs, we get a more informed picture.

The star Scheat co-setting with the Moon gives a positive energy to the process of projection, and strengthens the Moons' positive aspects with Jupiter and Venus. This paran with Scheat would be especially beneficial if you are working on a remedy for the muscles and the related activity of the muscles.

Jupiter is supported energetically by co-setting Bellatrix, which could relieve the stress of the negative opposing aspect with Venus. Antares co-culminating with Jupiter, is supportive to the immune system and the kidneys, also counteracting Jupiter's negative opposition to Venus.

Saturn, a negative planet presently in the positive sign of Aries, although not in aspect to the other planets, is co-culminating with Acturus and Capulus. These stars add stability to the normally disruptive Saturnic energies.

Antares co-rising, and Acturus co-setting, in paran with Mercury, can turn the reversible charge of the planet in a positive direction by adding clarity and focus.

Considering the weight of these stars directly affecting the aspects of this chart, we can deduce that this could be a good day to work. More assuredly, these stars would help on specific projects, for specific reasons. This is especially true if any of the stars in this chart are active in your birth chart, or the birth chart of the patient.

4. Star Paran Tables

From the book,

Star and Planet Combinations

by Bernadette Brady

ALCHEMY ASTROLOGY

40° North Latitude

Stars	Co-Culm	Position	Co-Culm	Position	Co-Rise 40N	Position	Co-Set 40N	Position
Alpheratz	Aries	02 17	Libra	02 17	Aquarius	18 37	Aries	23 50
Achernar	Aries	26 20	Libra	26 20	NotVis			
Capulus Nebula	Taurus	07 06	Scorpio	07 06	Cpole			
Menkar	Taurus	18 02	Scorpio	18 02	Gemini	02 40	Taurus	08 54
Bellatrix	Gemini	22 00	Saggitarius	22 00	Cancer	06 31	Gemini	08 47
El Nath	Gemini	22 16	Saggitarius	22 16	Gemini	16 06	Gemini	27 42
Sirius	Cancer	10 23	Capricorn	10 23	Leo	09 13	Gemini	08 50
Procyon	Cancer	23 00	Capricorn	23 00	Leo	04 55	Cancer	07 27
Acubens	Leo	12 10	Aquarius	12 10	Leo	15 59	Leo	05 47
Alphard	Leo	19 29	Aquarius	19 29	Virgo	05 30	Cancer	23 31
Regulus	Virgo	00 00	Pisces	00 00	Leo	29 39	Virgo	00 41
Alkes	Virgo	13 40	Pisces	13 40	Libra	00 50	Leo	11 06
Vindemiatrix	Libra	16 52	Aries	16 52	Libra	04 57	Scorpio	10 10
Spica	Libra	23 01	Aries	23 01	Libra	24 33	Libra	19 52
Thuban	Scorpio	03 19	Taurus	03 19	Cpole			
Arcturus	Scorpio	06 14	Taurus	06 14	Libra	13 31	Sagittarius	12 28
Antares	Saggitarius	09 03	Gemini	09 03	Sagittarius	13 12	Sagittarius	03 20
Ras Alhague	Saggitarius	24 14	Gemini	24 14	Scorpio	22 26	Capricorn	19 21
Facies Cluster	Capricorn	08 21	Cancer	08 21	Capricorn	09 07	Capricorn	07 40
Vega	Capricorn	08 29	Cancer	08 29	Scorpio	15 03	Aquarius	29 29
Sadalsuud	Aquarius	20 29	Leo	20 29	Aquarius	09 25	Aquarius	26 46
Deneb Algedi	Aquarius	24 27	Leo	24 27	Aquarius	28 18	Aquarius	22 27
Fomalhaut	Pisces	13 05	Virgo	13 05	Aries	21 44	Aquarius	24 58
Scheat	Pisces	14 44	Virgo	14 44	Capricorn	29 04	Aries	10 01

STAR PARAN TABLES

45° North Latitude

Stars	Co-Culm	Position	Co-Culm	Position	Co-Rise 45N	Position	Co-Set 45N	Position
Alpheratz	Aries	02 17	Libra	02 17	Aquarius	06 04	Aries	27 12
Achernar	Aries	26 20	Libra	26 20	NotVis			
Capulus Nebula	Taurus	07 06	Scorpio	07 06	Cpole			
Menkar	Taurus	18 02	Scorpio	18 02	Gemini	06 22	Taurus	07 33
Bellatrix	Gemini	22 00	Sagittarius	22 00	Cancer	06 31	Gemini	08 47
El Nath	Gemini	22 16	Sagittarius	22 16	Gemini	14 21	Gemini	29 02
Sirius	Cancer	10 23	Capricorn	10 23	Leo	13 45	Gemini	03 27
Procyon	Cancer	23 00	Capricorn	23 00	Leo	06 49	Cancer	04 06
Acubens	Leo	12 10	Aquarius	12 10	Leo	16 34	Leo	03 58
Alphard	Leo	19 29	Aquarius	19 29	Virgo	07 45	Cancer	17 28
Regulus	Virgo	00 00	Pisces	00 00	Leo	29 36	Virgo	00 55
Alkes	Virgo	13 40	Pisces	13 40	Libra	03 14	Leo	02 30
Vindemiatrix	Libra	16 52	Aries	16 52	Libra	03 20	Scorpio	16 49
Spica	Libra	23 01	Aries	23 01	Libra	24 46	Libra	18 48
Thuban	Scorpio	03 19	Taurus	03 19	Cpole			
Arcturus	Scorpio	06 14	Taurus	06 14	Libra	10 18	Sagittarius	20 21
Antares	Sagittarius	09 03	Gemini	09 03	Sagittarius	14 03	Sagittarius	01 35
Ras Alhague	Sagittarius	24 14	Gemini	24 14	Scorpio	23 31	Capricorn	26 52
Facies Cluster	Capricorn	08 21	Cancer	08 21	Capricorn	09 19	Capricorn	07 32
Vega	Capricorn	08 29	Cancer	08 29	Scorpio	04 37	Pisces	09 17
Sadalsuud	Aquarius	20 29	Leo	20 29	Aquarius	06 19	Aquarius	27 40
Deneb Algedi	Aquarius	24 27	Leo	24 27	Aquarius	29 36	Aquarius	22 09
Fomalhaut	Pisces	13 05	Virgo	13 05	Taurus	04 50	Aquarius	22 00
Scheat	Pisces	14 44	Virgo	14 44	Capricorn	18 05	Aries	13 49

ALCHEMY ASTROLOGY

50° North Latitude

Stars	Co-Culm	Position	Co-Culm	Position	Co-Rise 50N	Position	Co-Set 50N	Position
Alpheratz	Aries	02 17	Libra	02 17	Capricorn	20 16	Taurus	01 14
Achernar	Aries	26 20	Libra	26 20	NotVis			
Capulus Nebula	Taurus	07 06	Scorpio	07 06	Cpole			
Menkar	Taurus	18 02	Scorpio	18 02	Gemini	11 13	Taurus	06 09
Bellatrix	Gemini	22 00	Saggitarius	22 00	Cancer	12 41	Gemini	03 49
El Nath	Gemini	22 16	Saggitarius	22 16	Gemini	11 36	Cancer	00 56
Sirius	Cancer	10 23	Capricorn	10 23	Leo	18 44	Taurus	27 29
Procyon	Cancer	23 00	Capricorn	23 00	Leo	08 57	Gemini	29 59
Acubens	Leo	12 10	Aquarius	12 10	Leo	17 13	Leo	01 24
Alphard	Leo	19 29	Aquarius	19 29	Virgo	10 10	Cancer	10 05
Regulus	Virgo	00 00	Pisces	00 00	Leo	29 33	Virgo	01 18
Alkes	Virgo	13 40	Pisces	13 40	Libra	05 51	Cancer	21 42
Vindemiatrix	Libra	16 52	Aries	16 52	Libra	01 35	Scorpio	25 43
Spica	Libra	23 01	Aries	23 01	Libra	25 00	Libra	17 06
Thuban	Scorpio	03 19	Taurus	03 19	Cpole			
Arcturus	Scorpio	06 14	Taurus	06 14	Libra	06 46	Sagittarius	20 33
Antares	Saggitarius	09 03	Gemini	09 03	Sagittarius	15 10	Scorpio	28 47
Ras Alhague	Saggitarius	24 14	Gemini	24 14	Scorpio	18 49	Aquarius	02 06
Facies Cluster	Capricorn	08 21	Cancer	08 21	Capricorn	09 35	Capricorn	07 20
Vega	Capricorn	08 29	Cancer	08 29	Libra	18 37	Pisces	24 36
Sadalsuud	Aquarius	20 29	Leo	20 29	Aquarius	02 02	Aquarius	28 39
Deneb Algedi	Aquarius	24 27	Leo	24 27	Pisces	01 39	Aquarius	21 49
Fomalhaut	Pisces	13 05	Virgo	13 05	Taurus	23 00	Aquarius	18 22
Scheat	Pisces	14 44	Virgo	14 44	Capricorn	05 03	Aries	18 14

CHAPTER SEVEN – RELATING TO MATTER

1. Herbs To The Planets

Each day of the week is considered to be governed, or specifically influenced, by one of the planets. (See chart on page 34.) In the herbal work each plant is governed by at least one planet. There are several great resources in which you can research these signatures of the stars. *Culpepers' Color Herbal*, is the most well known book of this kind. Here are some common Herbs, and their associated planets, as listed in the *Alchemist's Handbook*:

Day	Governing Planet	Herb
Sunday	Sun	chamomile, rosemary, St. John's wort
Monday	Moon	chickweed, cleavers, willow tree
Tuesday	Mars	basil, garlic, nettle
Wednesday	Mercury	fennel, lavender, parsley
Thursday	Jupiter	dandelion, hyssop, roses
Friday	Venus	marshmallow, mugwort, self-heal
Saturday	Saturn	comfrey, horsetail, mullein

I was taught to make a simple alcohol tincture, or spagyric, of one herb for each planet. These are prepared on their associated day of the week, and time of day, when the general influences are positive. If you imbibe a few drops of the tincture, on their associated weekday, over time, it will bring about a purifying energy that will clean, and strengthen your body. These are called, "the seven basics." It is recommended that this be done for one year, which is one complete Earth orbit around the Sun, and a complete journey through all twelve signs of the zodiac to achieve the full benefit of the seven basics.

ALCHEMY ASTROLOGY

2. Metals To The Planets

Governing Planetary association with the metals;

Planet	Metal
Sun	Gold
Moon	Silver
Mercury	Mercury
Venus	Copper
Mars	Iron
Jupiter	Tin / Pewter
Saturn	Lead

3. Body To The Zodiac & Planets

When you decide which herb you want work with, look for associations with the area of the body you wish to remedy. For example, the Moon in Aries, in paran with Spica, between 03:25 and 06:51 on Monday, might be a great time to process a remedy for a disease like Alzheimers', or to improve memory, especially when the active influences are higher than the passive. The following charts list body associations with the zodiac, the planets, and the stars.

Zodiac	Body Association
Aries	head
Taurus	neck
Gemini	arms/shoulders
Cancer	chest
Leo	heart
Virgo	intestinal tract
Libra	kidneys
Scorpio	reproductive organs
Sagittarius	hips
Capricorn	thighs
Aquarius	legs
Pisces	feet

Planets	Body Association
Sun	heart
Moon	brain
Mercury	liver
Venus	veins
Mars	gall bladder
Jupiter	lungs
Saturn	spleen

4. Body To The Stars

Stars	Body Association
Achernar	Veins and arteries - circulatory
Acubens	Reproductive system
Alkes	Sex organs - genetic material
Alphard	Sex organs - libido
Alpheratz	Adrenal glands
Antares	Immune system
Arcturus	Knees and their cartilage sytem
Bellatrix	Shoulder bones
Capulus Nebula	Testosterone
Deneb Algedi	Knees and thighs
El Nath	Male genitalia
Facies Cluster	Eyes - vision quality
Fomalhaut	Histamine reactions
Menkar	Olfactory faculty - smell
Procyon	Physical fitness
Ras Alhague	Lymphatic system
Regulus	Heart and vascular system
Sadalsuud	Legs and ankles
Scheat	Motor skill and muscle memory
Sirius	Sweat glands
Spica	Memory
Thuban	Nutrient storage in the body
Vega	Sleep
Vindemiatrix	Kidneys

ALCHEMY ASTROLOGY

CHAPTER EIGHT - 2012
The Galactic Alignment

December 21, 2012, doesn't signify a specific date for disaster. To the Mayans it unambiguously marked the end of one of their cyclic calendars denoting a general time of rebirth for mankind. It is simply coincidental that other astronomical influences may be in evidence for heralding a series of minor catastrophes at this time.

The Milky Way Galaxy is a disc, like our solar system. The plane, or disc, emits radiation of several wavelengths, and, as our planet becomes centered on these energies, they are focused with a greater intensity. This is similar to aligning a magnifying glass perpendicular to the rays of the Sun. Combined with certain positions of the planets and the Moon in relation to the Sun, a triggering type of stress is fixed on the Earth's tectonic plates. (Russian Institute of Physics of the Earth, 2011)

On December 21, 1998, our Sun, as viewed from our planet, aligned with the black hole at the galactic center. In the last forty years, as we approached this occurrence, earthquakes, volcanoes, and severe weather events, have been on the increase and are continuing to this day. Presently, Jupiter and Saturn are pulling the Earth and Sun in opposite directions. Add in the galactic center energies and you have a dynamic situation. The Sun's solar activity is also amplified by Saturn and Jupiter, and we know that the resulting solar flares affect the Earth's magnetosphere and weather. Scientists have determined that the solar wind alone, through its electrical and magnetic influences, create the auroras. It's reasonable to speculate that the Earth is also influenced by changes in flare boosted solar winds in ways not yet confirmed by modern science. I surmise, that the gravity of the planets, like obstacles in a stream, potentise the solar currents and I see evidence that this theory is on the verge of being proven.

On December 21st, 2012, the Sun, and Earth, will roughly be in alignment nearest to the center of our Milky Way Galaxy, as they do every year, focusing all the energies entering our solar system, onto the Earth. The Moon will move through this line in the days following. Similar to wind currents that are divided by the mountains in Hawai'i, which allow canoes to be navigated against strong predominating oceanic winds for up to 3,000 kilometers (Schmidt, 2003), this focused energy will cause a back draft that could pull in any deep space objects in this trough towards our planet. It is not certain that this alignment will cause a single catastrophic event, but it is not unlikely that an increase in volcanic eruptions and earthquakes may occur (HEWS, 2010). Clouds of volcanic dust and ash can block the Sun and cause an extended winter season that could devastate crops and other food supplies.

In conclusion, in 2012, the Earth will be approximately 6° past alignment with the black hole at the center of our Milky Way galaxy. Gradually, and as Jupiter and Saturn pass opposition, things will improve.

ALCHEMY ASTROLOGY

Our next challenge will be on July 6, 2020, when Jupiter and Saturn are conjunct, aligned with the Moon and Earth opposite the Sun, and in the general vicinity of the Galactic Center.

The following chart shows the positions of the planets in their orbits. This kind of chart can help you visualize the energy paths and aspects between the Earth and the planets. Earthquakes and dates when alchemists transmuted gold in front of credible witnesses (Kollerstrom) are of particular interest when learning to deduce charting information. More examples are on my website (Wilkerson, 2010).

2012

GREENWICH 12-21-2012   12:00

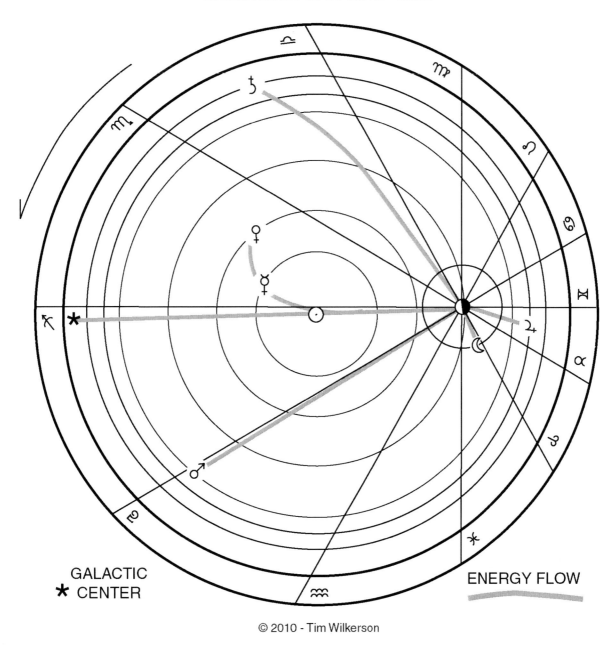

GALACTIC
★ CENTER

ENERGY FLOW

© 2010 - Tim Wilkerson

57

ALCHEMY ASTROLOGY

The top chart is a witnessed gold making date. The bottom chart is an earthquake date. In both charts Jupiter, Saturn, and Mercury are in similar positions. A comparison of these events is evidence of astronomical influences on our Earth.

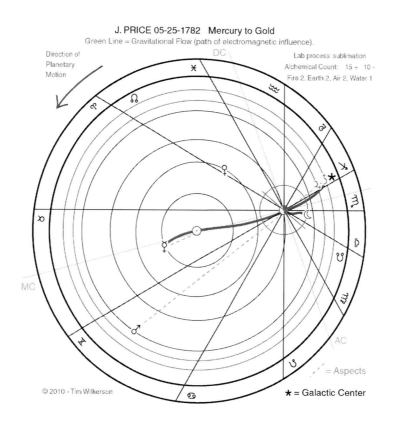

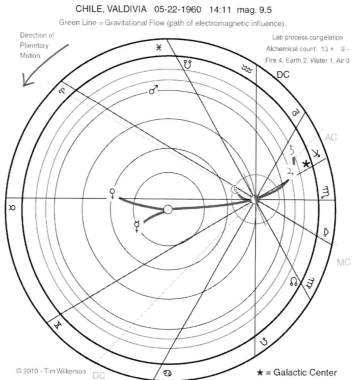

ALCHEMY ASTROLOGY

Appendix A

1. Planets – Scientific Stats

Planet	Gravity	Rotation	Satelites	Orbit	Distance
Mercury	0.38	59.00 days	0	88.00 days	57.9 Mkm
Venus	0.9	243.00 days	0	224.70 days	108.2 Mkm
Earth	1	23.93 hrs.	1	365.25 days	149.6 Mkm
Moon	0.05	27.32 hrs.	1	27.32 days	384,400 km
Mars	0.38	24.63 hrs.	2	687.00 days	227.9 Mkm
Jupiter	2.53	9.93 hrs.	63+	11.86 years	778.3 Mkm
Saturn	1.07	10.06 hrs.	56+	29.46 years	1.43 Bkm

2. Stars – Date of Record, Distance, & Type

Stars	Date	Distance	Type
	of record	in light years	
Achernar	3,500 bc	144	Blue-White main sequence
Acubens	4,000 bc	174	White
Alkes	3,500 bc	174	Orange Giant
Alphard	1,200 bc	177	Orange Giant
Alpheratz	6,000 bc	97	Blue-white dark companion
Antares	3,000 bc	604	Red Supergiant/Blue Companion
Arcturus	6,000 bc	37	Orange Giant
Bellatrix	4,667 bc	243	Blue Giant
Capulus Nebula	1,433 bc	3,500	Planetary Nebulae
Deneb Algedi	3,000 bc	109	Yellow Super Giant / Yellow Giant, both double
El Nath	5,000 bc	131	Blue-White Giant
Facies Cluster	6,000 bc	10,000	Globular cluster
Fomalhaut	3,500 bc	25	Blue-White main sequence
Menkar	2,000 bc	220	Red Giant/Blue White companion
Procyon	2,400 bc	11.4	White/White Dwarf companion
Ras Alhague	3,500 bc	47	White main sequence
Regulus	2,400 bc	77	Blue-White main sequence
Sadalsuud	1,300 bc	760	Yellow Super Giant
Scheat	4,000 bc	199	Red Giant
Sirius	58,000 bc	8.6	White/White Dwarf companion
Spica	3,200 bc	262	Blue-White main sequence
Thuban	2,790 bc	309	Blue-White Giant
Vega	12,000 bc	25	Blue-White main sequence
Vindemiatrix	3,500 bc	102	Yellow Giant

3. Stars – Names & Scientific Spectral Type

Star	Spectral Type	Spectral Type +
Achernar	B	B3Vpe
Acubens	A	A5m
Alkes	K	K0+III
Alphard	K	K3II-III
Alpheratz	B	B8IVpMnHg
Antares	M	M1Ib + B2.5V
Arcturus	K	K1.5IIIFe-0.5
Bellatrix	B	B2III
Capulus Nebula		
Deneb Algedi	A	Am
El Nath	B	B7III
Facies Cluster		
Fomalhaut	A	A3V
Menkar	M	M1.5IIIa
Procyon	F	F5IV-V
Ras Alhague	A	A5III
Regulus	B	B7V
Sadalsuud	G	G0Ib
Scheat	M	M2.5II-III
Sirius	A	A1Vm
Spica	B	B1III-IV+B2V
Thuban	A	A0III
Vega	A	A0Va
Vindemiatrix	G	G8IIIab

Appendix A

4. Stars – Spectral Type & Associated Temperature Ranges

The cooler the temperature, the more varied the type of metals that are present as detected by spectral analysis. An M2 star such as Scheat, when in paran to a planet, may influence work with metals, minerals, and plants with high levels of organic metals, more than a type B star, such as Spica.

Spectral Letter	Temperature (F)	Temperature (C)
O	more than 37,000	more than 20,500
B	17,000 - 37,000	9,430 - 20,500
A	12,500 - 17,000	6,930 - 9,430
F	10,300 - 12,500	5,700 - 6,930
G	8,000 - 10,300	4,400 - 5,700
K	5,500 - 8,000	3,040 - 4,400
M	less than 5,500	less than 3,040

Spectral Letter / Number	Temperature (K)
O	35,000
B2	22,000
B5	16,400
A0	10,800
A5	8,600
F0	7,200
F5	6,500
G0	5,900
G5	5,600
K0	5,200
K5	4,400
M0	3,700
M2	3,500

5. Constellations Of Stars

The following chart lists the constellation each star appears in. Each stellar grouping has a different traditional story depending upon which society the tale originates. You may wish to look deeper into other star lore information based on these associations.

Stars	Constellation
Achernar	Eridanus - River
Acubens	Cancer - Crab
Alkes	Crater - Cup
Alphard	Hydra - Water Snake
Alpheratz	Andromeda - Chained Lady
Antares	Scorpius - Scorpion
Arcturus	Bootes - Herdsman
Bellatrix	Orion - Hunter
Capulus Nebula	Perseus - Warrior/Champion
Deneb Algedi	Capricorn - Sea Goat
El Nath	Taurus - Bull
Facies Cluster	Sagittarius - Archer
Fomalhaut	Piscis Australis - Southern Fish
Menkar	Cetus - Whale
Procyon	Canis Minor - Lesser Dog
Ras Alhague	Ophiuchus - Serpent Holder
Regulus	Leo - Lion
Sadalsuud	Aquarius - Water Pourer
Scheat	Pegasus - Flying Horse
Sirius	Canis Major - Greater Dog
Spica	Virgo - Virgin
Thuban	Draco - Dragon
Vega	Lyra - Lyre/Vulture

Appendix B

You, as the purchaser/owner of this book, are free to copy the following for your own personal use, as a professional teaching aid, or for use in a laboratory research environment.

Commercial use of the following charts is by permission from the author only.

Files are available online at: alchemyastrology.com.

1. Blank Astrology Chart

This blank chart doesn't include the zodiac signs so that you may choose which sign to put in the first position.

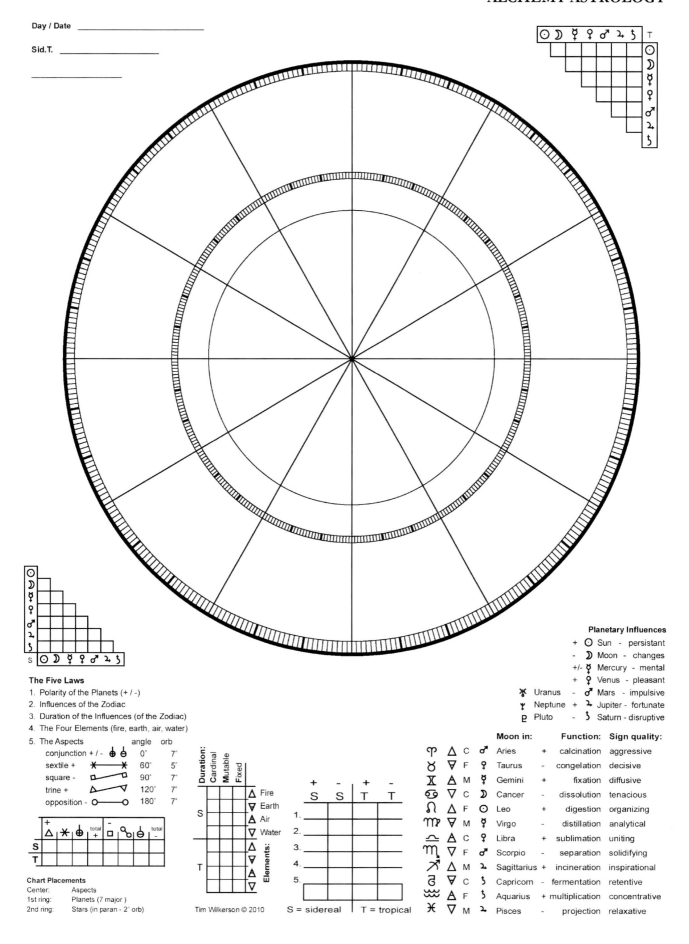

2. Transparency

Print this chart on a transparent plastic sheet for use in converting tropical astrology to sidereal astrology. Special printable sheets, specific to your type of copy machine, or printer, can be purchased at an office supply store.

ALCHEMY ASTROLOGY

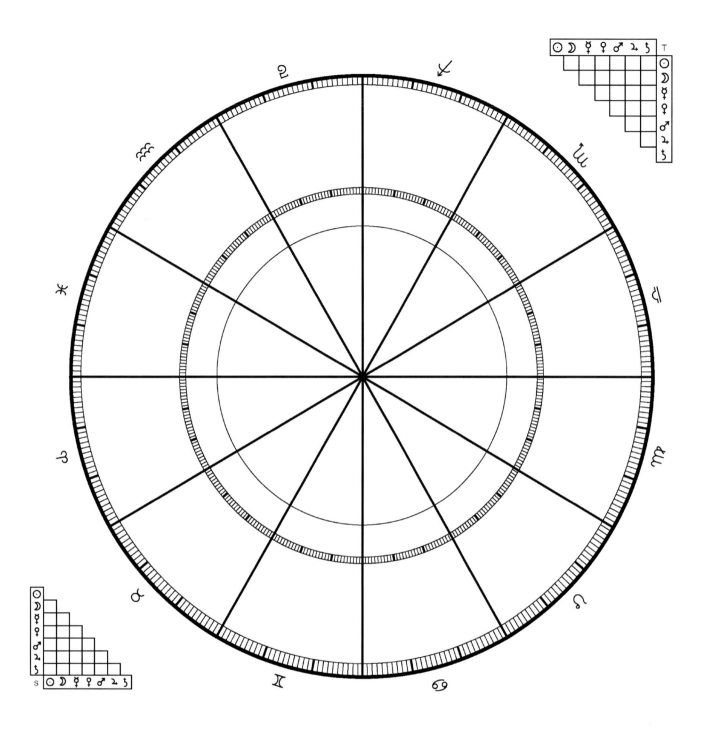

ALCHEMY ASTROLOGY

3. Blank Earth Centered Orbital Charts

How to add information to these charts:

a. Choose the chart that most closely represents the degree of the Sun as listed in the ephemeris.
b. Draw the appropriate zodiac sign next to the 30° section where the Sun is located, and continue counter clock wise around the wheel, until you have drawn all the signs in order. (You can use the blank transparency chart in this appendix as a reference.)
c. Place the planets at their assigned location (as listed in the Ephemeris), on their correct circular orbits. Start nearest to the Sun with Mercury, and continue in this order: Venus, (Earth is already in place), Mars, Jupiter, and Saturn.
d. The circle around the Earth is the orbit of the Moon, draw its symbol in the assigned zodiac sign.

© 2011 - Tim Wilkerson

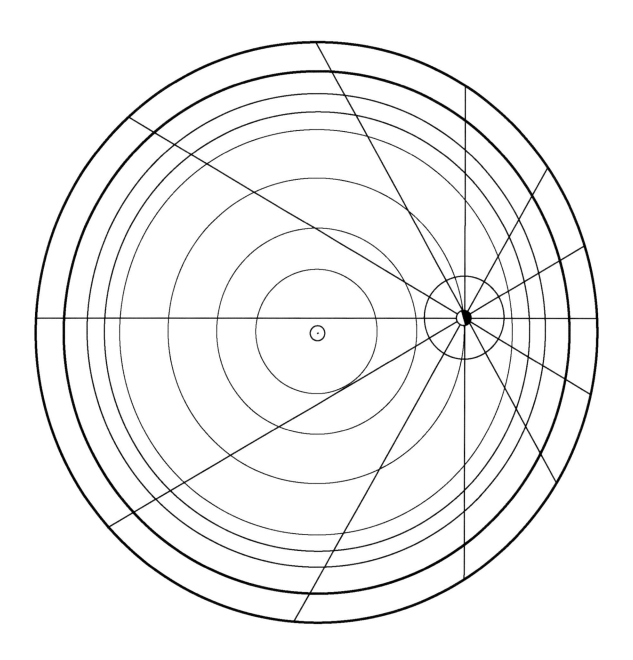

ALCHEMY ASTROLOGY

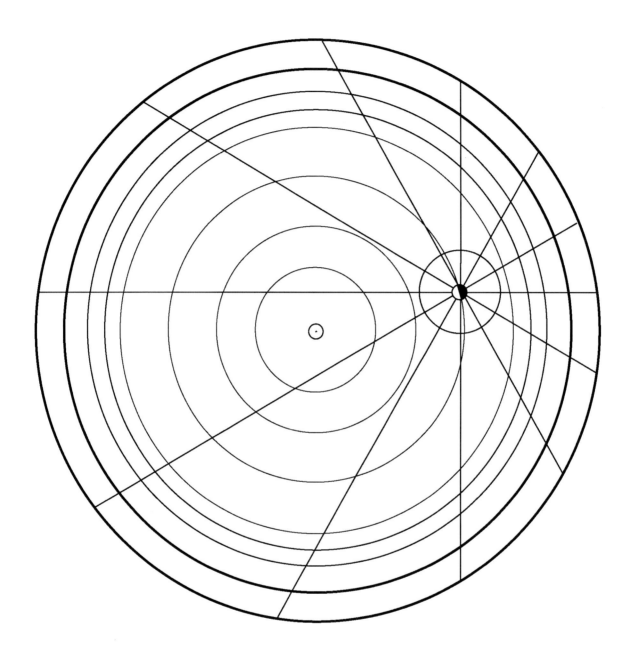

ALCHEMY ASTROLOGY

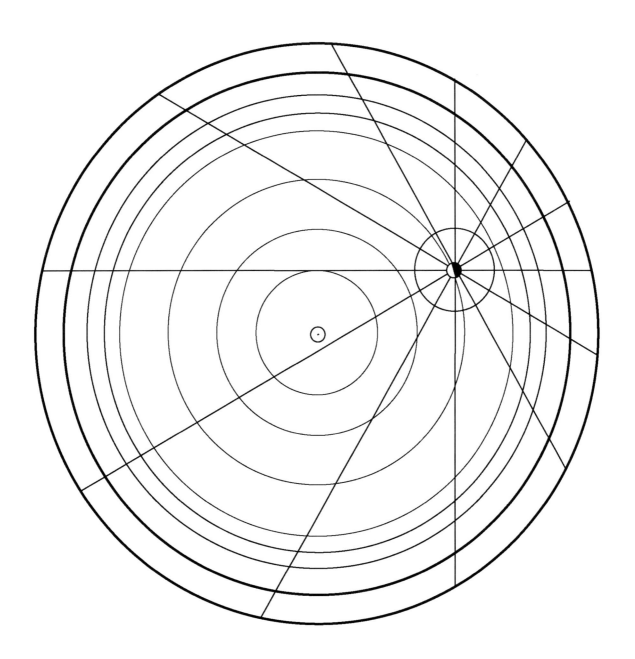

4. Basic Lab Form

ALCHEMY ASTROLOGY

	DATE / DAY	NAME	PLANETARY INFLUENCE
TIME			

5. Quick Chart – Four Day

Use this chart to quickly plan future experiments. The influences predicted in a chart can remain the same for approximately three days. Look ahead in the ephemeris for successive dates when the Moon, or one of the planets, change zodiac signs. Compute the Five Laws to locate days with more active/positive influences.

ALCHEMY ASTROLOGY

Day / Date _____
Sid T. _____

Day / Date _____
Sid T. _____

	+	-
1.		
2.		
3.		
4.		
5.		

	+	-
△		□
✶		⚭
⊕		⊖

	+	-
1.		
2.		
3.		
4.		
5.		

	+	-
△		□
✶		⚭
⊕		⊖

Day / Date _____
Sid T. _____

Day / Date _____
Sid T. _____

	+	-
1.		
2.		
3.		
4.		
5.		

	+	-
△		□
✶		⚭
⊕		⊖

	+	-
1.		
2.		
3.		
4.		
5.		

	+	-
△		□
✶		⚭
⊕		⊖

© 2011 - Tim Wilkerson

6. Salts - Calcining Grayscale Reference

Match the ash/salt shade to this reference chart, and record the single digit number, and the time of your observation, in your lab notes. Include observations of visible color hues.

Derived from basic html grayscale codes used on the internet and listed in Row 3.

Black	1	2	3	4	5	6	7
#000000	#111111	#222222	#333333	#444444	#555555	#666666	#777777

8	9	10	11	12	13	14	White
#888888	#999999	#AAAAAA	#BBBBBB	#CCCCCC	#DDDDDD	#EEEEEE	#FFFFFF

© 2011 - Tim Wilkerson

Appendix C

1. List of Website Resources for Further Study

a. Alchemy

AlchemyAstrology: http://www.alchemyastrology.com

Alchemy Guild - Alchemy Information: http://www.alchemyguild.memberlodge.org/

Alchemy Study: http://www.alchemystudy.com/

Dennis William Hauck - The Red Phase: http://www.dwhauck.com/

Gary Stadler – Alchemy glassware: http://heartmagic.com/EssentialDistiller.html

Robert Allen Bartlett - classes: http://realalchemy.org/default.aspx

b. Astronomy

Articles:

An introduction to Gravity, and Einstein's General Theory of Relativity, contains graphic examples: http://www4.ncsu.edu/unity/lockers/users/f/felder/public/kenny/papers/gr1.html

Gravity, for those who want to know more, contains graphic examples. Author, Matt_McIrvin:

Astronomy Websites of Interest:

Audio - Hear intriguing radio waves that NASA's Cassini spacecraft collected near Jupiter in January 2001: http://www.nasa.gov/wav/52858main_20010104-ia-a.wav

University of California, San Diego: http://cassfos02.ucsd.edu/public/astroed.html#TUTORIAL

Video - Visual astronomical explanation of the Galactic Alignment on December 21, 2012: http://www.youtube.com/watch?v=gc_nXcPuT1Q

c. Astrology

Astrology Basics: (World Mysteries, 2010): http://www.world-mysteries.com/amr/astrology_basics.htm

Ephemeris: http://www.astro.com/swisseph/swepha_e.htm.

d. Documentaries

The Lives of Extremophiles:
http://www.pbs.org/wgbh/nova/caves/extremophiles.html\

Tales from the Underground, by David Wolfe:
http://www.youtube.com/watch?v=-ySxHKiy9wY

Inside The Milky Way
http://channel.nationalgeographic.com/episode/inside-the-milky-way-4605/Overview

e. Microbial Research

"Sediment Microbial Community Composition and Methylmercury," K. M. Batten and K. M. Scow:
http://www.springerlink.com/content/ahwq84fqmyhfxn09/

"The Lives of Extremophiles," *NOVA*, article.
http://www.pbs.org/wgbh/nova/caves/extremophiles.html

2. Recommended Books

Alchemy, Brian Cotnoir, Pub. Weiser. © 2006.

Alchemical Guide to Herbs and Food, Dennis William Hauch. Alchemy Guild, © 2005.

Practical Handbook of Plant Alchemy, Manfred M. Junius, Pub. Inner Traditons, © 1979.

Real Alchemy, Robert Alan Bartlett. Pub. Quinguangle Press, © 2006.

Stars and Planets, Princeton Field Guides, Ian Ridpath and Wil Tirion, Pub. Princeton University Press, © 2007.

Star and Planet Combinations, Bernadette Brady, Pub. The Wessex Astrologer, © 2008.

Star Lore, Myths, Legends, and Facts, William Tyler Olcott. Pub. Dover Publications, © 2004.

The Alchemist's Handbook, Frater Albertus, Pub. Samuel Weiser, Inc., © 1974.

The Only Way To Learn Astrology, Volumn 1, Marion D. March and Joan McEvers, Pub ACS, © 1976.

The Portal, Russell Burton House, Pub. Triad Publishing,© 2007

The Way of the Crucible, Robert Alan Bartlett. Pub. Lulu, © 2008.

Works Cited

Advameg, I. (n.d.). *Harnessing Invisible Power*. Retrieved 2010 from Science Clarified: http://www.scienceclarified.com/scitech/Bacteria-and-Viruses/Harnessing-Invisible-Power.html

Albertus, F. (1974). *Alchemist's Handbook*. York Beach, Maine, U.S.A.: Samuel Weiser, Inc.

Amherst. (2001 йил 10-September). *'Gold Bug' Sheds Light on How Some Gold Deposits Formed*. Retrieved 2010 from Space Daily - Your Portal To Space: http://www.spacedaily.com/news/early-earth-01j.html

Anil Bhardwaj, R. F. (2005 йил 10-May). *Chandra Observation of an X-Ray Flare at Saturn: Evidence of Direct Solar Control on Saturn's Disk X-Ray Emissions*. Retrieved 2010 from The Astrophysical Journal: http://iopscience.iop.org/1538-4357/624/2/L121/19386.text.html

Bartlett, R. A. (2006). *Real Alchemy, A Primer of Practical Alchemy*. U.S.A.: Quinquangle Press.

Bartlett, R. A. (2008). *The Way of the Crucible*. U.S.A.: Lulu.

BBC. (2010). *Louis Pasteur*. Retrieved 2010 from BBC Historic Figures: http://www.bbc.co.uk/history/historic_figures/pasteur_louis.shtml

Blair-Ewart, A. (1999). *Sidereal Astrology Originated in Ancient Egypt*. Retrieved 2010 from western sidereal astrology: http://westernsiderealastrology.wordpress.com/sidereal-astrology-originated-in-ancient-egypt/

Bortman, H. (2005 йил 17-November). *SETI Sets its Sights on M Dwarfs*. Retrieved 2010 from Astrobiology Magazine Cosmic Evolution: http://www.astrobio.net/exclusive/1776/seti-sets-its-sights-on-m-dwarfs

Boyd, R. S. (2009 йил 14-July). *Gravity wells could provide 'parking lots' for spaceships*. Retrieved 2010 йил 11-October from McClatchy: http://www.mcclatchydc.com/2009/07/14/71824/gravity-wells-could-provide-parking.html

Bucher, J. (2005). *Fishing and Moon Phases*. Retrieved 2010 from Angler Guide: http://www.anglerguide.com/articles/522.html

Bulgerin, M. (2002 йил 14-February). *Bitstream Net*. Retrieved 2010 from Geometry of Eclipses: http://members.bitstream.net/~bunlion/bpi/EclGeom.html

Chandra. (2006 йил 21-August). *Gravitational Lensing Explanation*. Retrieved 2010 from Chandra X-Ray Observatory: http://chandra.harvard.edu/photo/2006/1e0657/more.html

Chandra. (2006 йил 21-August). *NASA Finds Direct Proof of Dark Matter*. Retrieved 2010 from Chandra X-Ray Observatory: http://chandra.harvard.edu/photo/2006/1e0657/index.html

Degrandi-Hoffman. (2009 йил 1-August). *The Importance of Microbes in Nutrition and Health of Honey Bee Colonies*. Retrieved 2010 from USDA: http://www.ars.usda.gov/research/publications/publications.htm?seq_no_115=242630

Fourmilab. (2010). *New Moon, Full Moon, Apogee, Perigee*. Retrieved 2010 from Fourmilab: http://www.fourmilab.ch/earthview/pacalc.html

Geographic, N. (Writer). (2010). *Inside The Milky Way* [Motion Picture].

Gill, V. (2010 йил 14-July). *Plants 'can think and remember'*. Retrieved 2010 from BBC News: http://www.bbc.co.uk/news/10598926

Goudarzi, S. (2007 йил 8-October). *Plants Communicate to Warn Against Danger*. Retrieved 2010 from LiveScience: http://www.livescience.com/animals/071008-plants-communicate.html

Gregory Cajete, P. (2000). *Native Science, The Natural Laws of Interdependence*. Clear Light Publications.

Guinard, P. (1999). *Cyril Fagan. The Oktotopos, an ancient astrological 8-fold division of space*. Retrieved 2010 from C.U.R.A. The International Astrology Research Center : http://cura.free.fr/docum/02fagan.html

Hansford, D. (2007 йил 22-October). *Moonlight Triggers Mass Coral "Romance"*. Retrieved 2010 from National Geographic News: http://news.nationalgeographic.com/news/2007/10/071019-coral-spawning.html

Hanson, B. (1995 йил 6-October). *They're alive! - bacterial spores survive for possibly millions of years*. Retrieved 2010 from Science World: http://findarticles.com/p/articles/mi_m1590/is_n3_v52/ai_17425124/

Hatch, R. A. (1998). *Isaac Newton Biography - Newton's Life, Career, Work*. Retrieved 2010 from University of Florida: http://www.clas.ufl.edu/users/ufhatch/pages/01-courses/current-courses/08sr-newton.htm

Hauck, D. (n.d.). *Isaac Newton the Alchemist*. Retrieved 2010 from Alchemylab: http://www.alchemylab.com/isaac_newton.htm

Hauck, D. W. (2008). *Complete Idiot's Guide to Alchemy*. (N. Wagner, Ed.) New York, New York, U.S.A.: Marie Butler-Knight.

HEWS. (2010). *Current Earthquake List*. Retrieved 2010 from Humanitarian Early Warning Service: http://www.hewsweb.org/seismic/

HMPDACC. (2010). *Data Analysis and Coordination Center*. Retrieved 2010 from Human Microbiome Project DACC: http://www.hmpdacc.org/

Hollocher, K. (2007). *Moon Rocks Background Material*. Retrieved 2010 from Union College: http://minerva.union.edu/hollochk/c_petrology/moon_rocks/background.htm

JPL NASA. (2010). *Voyager The Interstellar Mission*. Retrieved 2010 from Jet Propulsion Laboratory California Institute of Technology: http://voyager.jpl.nasa.gov/multimedia/flash_html.html

Junius, M. M. (1985). *Practical Handbook of Plant Alchemy, How to Prepare Medicinal Essence, Tinctures and Elixirs*. (L. Muller, Trans.) New York, New York, U.S.A.: Inner Traditions International, Ltd.

Klyce, B. (n.d.). *Bacteria: The Space Colonists*. Retrieved 2010 from Cosmic Ancestry: http://www.panspermia.org/bacteria.htm

Kollerstrom, N. (n.d.). *Golden Moments - Nick Kollerstrom*. Retrieved 2010 from The Alchemy Website: http://www.alchemywebsite.com/golden_m.html
Lake-Link. (2010). *Peak Fishing Times & Lunar Phases*. Retrieved 2010 from Lake-Link: http://www.lake-link.com/moon/
Lin, R. P., Mitchell, D. L., Curtis, D. W., Anderson, K. A., Carlson, C. W., McFadden, J., et al. (1998 йил 4-September). *Lunar Surface Magnetic Fields and Their Interaction with the Solar Wind: Results from Lunar Prospector*. Retrieved 2010 from Science / AAAS - Science Magazine: Research, news and commentary: http://www.sciencemag.org/cgi/content/full/281/5382/1480
Lunarium. (2010). *Lunar Gardening and its Different Traditions*. Retrieved 2010 from Lunarium: http://www.lunarium.co.uk/astroschool/articles/lunar-gardening-4.jsp
Magda Havas, P. (2009 йил 13-October). *Power Quality/Dirty Electricity*. Retrieved 2010 from Environmental Studies Research Papers: http://www.magdahavas.org/tag/dirty-electricity/
McEvers, J., & March, M. D. (1976). *The Only Way To Learn Astrology* (Revised edition ©1981 ACS Publications ed., Vol. 1). San Diego, California, U.S.A.: Astro-Analytic Publications.
merlynne6. (2009). *What the Lascaux Cave Paintings Tell Us About How Our Ancestors Understood the Stars*. Retrieved 2010 from Environmental graffiti: http://www.environmentalgraffiti.com/sciencetech/what-the-lascaux-cave-paintings-tell-us-about-how-our-ancestors-understood-the-stars/15506
Michelsoen, N. F., & Pottenger, R. (1997). *The American Ephemeris for the 21st Century, 2000 – 2050*. ACS Publications.
Mora, C. (2010 йил 20-May). *Human Microbiome Project publishes the first genomic collection of human microbes*. Retrieved 2010 from Wireupdate: http://wireupdate.com/wires/5299/human-microbiome-project-publishes-the-first-genomic-collection-of-human-microbes/
Mosher, D. (2010 йил 26-October). *Signs of Destroyed Dark Matter Found in Milky Way's Core*. Retrieved 2010 from Wired Science: http://www.wired.com/wiredscience/2010/10/dark-matter-milky-way/
Mountain Waves Healing Arts. (2009 йил 18-February). *Good Vibrations Sand Example*. Retrieved 2010 from YouTube: http://www.youtube.com/watch?v=OKHBPTKZ9Kc&NR=1
NASA. (2000 йил 13-January). *Panspermia theoretically possible, say scientists*. Retrieved 2010 from NASA - Astrobiology - Latest News: http://astrobiology.arc.nasa.gov/news/expandnews.cfm?id=295
NatGeo. (2007 йил 9-April). *Giant Crystal Cave Comes to Light*. Retrieved 2010 from National Geographic News: http://news.nationalgeographic.com/news/2007/04/photogalleries/giant-crystals-cave/
NIH. (n.d.). *Human Microbiome Project Overview*. Retrieved 2010 from National Institute of Health: http://nihroadmap.nih.gov/hmp/
NOAA. (2010). *Space Weather Prediction Center*. Retrieved 2010 from NOAA: http://www.swpc.noaa.gov/
Olcott, W. T. (2004). *Star Lore, Myths, Legends, and Facts*. Dover Publications.
Parkes, R. J. (n.d.). *Geomicrobiology - Evolution of bacteria: stromatolites, photosynthesis, Bacteria and ore deposits: 'black smokers'*. Retrieved 2010 from science.jrank.org: http://science.jrank.org/pages/47644/geomicrobiology.html
Rappenglueck, M. A. (2006). *The Planet Earth: Carved and Drawn Prehistoric Maps of the Cosmos*. Retrieved 2010 from Space Today Online, Solar System Planet Earth, Ancient Astronomy: http://www.spacetoday.org/SolSys/Earth/OldStarCharts.html
Ridpath, I., & Tirion, W. (2007). *Stars and Planets*. Princeton Field Guides.
Sanders, R. (2009 йил 23-December). *Sun and moon trigger deep tremors on San Andreas Fault*. Retrieved 2010 from UC Berkeley News: http://berkeley.edu/news/media/releases/2009/12/23_Parkfieldtremors.shtml
Schmidt, L. J. (2003 йил 2-October). *Little Islands, Big Wake*. Retrieved 2010 from NASA Earth Observatory: http://earthobservatory.nasa.gov/Features/Wake/
Singer, E. (2007 йил 2-May). *The Next Human Genome Project: Our Microbes*. Retrieved 2010 from Technology Review: http://www.technologyreview.com/biotech/18618/
SIUC. (2010). *Shaken and Shocked*. Retrieved 2010 from Perspectives, Southern Illinois University Carbondale: http://perspect.siuc.edu/05_fall/earthquakes.html
Smithsonian Institute. (2010). *Volcanoes of the World*. Retrieved 2010 from Global Vulcanism Program: http://www.volcano.si.edu/world/find_eruptions.cfm
SOHO. (2010). *The Very Lastest SOHO Images*. Retrieved 2010 from SOHO: http://sohowww.nascom.nasa.gov/data/realtime-images.html
Staelin, E. (2006 йил March). *Strong Bones or Osteoporosis, Part I: Beware of Too Much Calcium*. Retrieved 2010 from Healing Teeth Naturally: http://www.healingteethnaturally.com/biological-transmutation-calcium-from-horsetail-silica.html
STEREO. (n.d.). *Space Weather*. Retrieved 2010 from STEREO: http://stereo.gsfc.nasa.gov/img/spaceweather/movies/BarMagC.mov
Than, K. (2007 йил 12-March). *Sun Blamed for Warming of Earth and Other Worlds*. Retrieved 2010 from LIVE SCIENCE: http://www.livescience.com/environment/070312_solarsys_warming.html
Tyson, P. (2000). *Into the Abyss | Living at Extremes*. Retrieved 2010 from NOVA: http://www.pbs.org/wgbh/nova/abyss/life/extremes.html
Tyson, P., & Northup, D. (n.d.). *Mysterious Life of Caves, The Lives of Extremophiles*. Retrieved 2010 from NOVA Science Programming On Air and Online: http://www.pbs.org/wgbh/nova/caves/extremophiles.html
U of C. (n.d.). *LIGHT AND LIGHTING FOR POULTRY*. Retrieved 2010 from University of Connecticut, Information Technology Services: http://www.sp.uconn.edu/~mdarre/poultrypages/light_inset.html

USGS. (2010). *United States Geological Survey, Earthquake Stats.* Retrieved 2010 from USGS: http://earthquake.usgs.gov/earthquakes/eqarchives/

USNO. (2010). *Complete Sun and Moon Data for One Day: U.S. Cities and Towns.* Retrieved 2010 from Naval Oceanography Portal: http://www.usno.navy.mil/USNO/astronomical-applications/data-services/rs-one-day-us

White, M. (2010). *Dark Matter.* Retrieved 2010 from UC Berkeley Astronomy Department: http://astro.berkeley.edu/~mwhite/darkmatter/dm.html

Wilkerson, T. (2010). *2012.* Retrieved 2010 from Alchemy Astrology: http://www.risingstarmusic.com/alchemyastrology/2012.html

World Mysteries. (2010). *Astrology and Alchemy Basics.* Retrieved 2010 from World-Mysteries: http://www.world-mysteries.com/amr/astrology_basics.htm